Frontiers in Approximation Theory

SERIES ON CONCRETE AND APPLICABLE MATHEMATICS

ISSN: 1793-1142

Series Editor: Professor George A. Anastassiou
Department of Mathematical Sciences
University of Memphis
Memphis, TN 38152, USA

---

*Published**

Vol. 6 Topics on Stability and Periodicity in Abstract Differential Equations
by James H. Liu, Gaston M. N'Guérékata & Nguyen Van Minh

Vol. 7 Probabilistic Inequalities
by George A. Anastassiou

Vol. 8 Approximation by Complex Bernstein and Convolution Type Operators
by Sorin G. Gal

Vol. 9 Distribution Theory and Applications
by Abdellah El Kinani & Mohamed Oudadess

Vol. 10 Theory and Examples of Ordinary Differential Equations
by Chin-Yuan Lin

Vol. 11 Advanced Inequalities
by George A. Anastassiou

Vol. 12 Markov Processes, Feller Semigroups and Evolution Equations
by Jan A. van Casteren

Vol. 13 Problems in Probability, Second Edition
by T. M. Mills

Vol. 14 Evolution Equations with a Complex Spatial Variable
by Ciprian G. Gal, Sorin G. Gal & Jerome A. Goldstein

Vol. 15 An Exponential Function Approach to Parabolic Equations
by Chin-Yuan Lin

Vol. 16 Frontiers in Approximation Theory
by George A. Anastassiou

*To view the complete list of the published volumes in the series, please visit:
http://www.worldscientific/series/scaam

Series on Concrete and Applicable Mathematics – Vol. 16

Frontiers in Approximation Theory

George A. Anastassiou

University of Memphis, USA

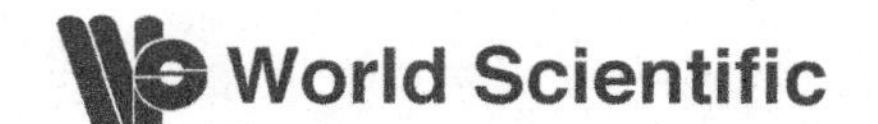

NEW JERSEY • LONDON • SINGAPORE • BEIJING • SHANGHAI • HONG KONG • TAIPEI • CHENNAI

Published by

World Scientific Publishing Co. Pte. Ltd.

5 Toh Tuck Link, Singapore 596224

USA office: 27 Warren Street, Suite 401-402, Hackensack, NJ 07601

UK office: 57 Shelton Street, Covent Garden, London WC2H 9HE

Library of Congress Cataloging-in-Publication Data

Anastassiou, George A., 1952–

Frontiers in approximation theory / George Anastassiou (University of Memphis, USA).

pages cm. -- (Series on concrete and applicable mathematics (SCAM) ; vol. 16)

Includes bibliographical references and index.

ISBN 978-9814696081 (alk. paper)

1. Approximation theory. 2. Monotone operators. 3. Fractional differential equations. I. Title.

QA221.A5363 2015

511'.4--dc23

2015018610

British Library Cataloguing-in-Publication Data

A catalogue record for this book is available from the British Library.

In-house Editors: Kwong Lai Fun/V. Vishnu Mohan

Typeset by Stallion Press

Email: enquiries@stallionpress.com

Printed in Singapore

"The Main feature of life on this Planet is continuation. That is continuous movement of everything which involves stops, gaps, and jumps."

Preface

In this monograph we present recent work of last five years of the author in Approximation Theory. It is the natural outgrowth of his related publications. Chapters are self-contained and advanced courses can be taught out of this book. An extensive list of references is given per chapter.

The topics covered are diverse. The first eight chapters are dedicated to fractional monotone approximation theory introduced for the first time by the author, taking the related ordinary theory of usual differentiation at the fractional differentiation level having polynomials and splines as approximators. Very little is written so far about fractional approximation theory which is at its infancy. Chapters 9–10 are dedicated to the approximation by discrete singular operators of Favard style, e.g. of Picard and Gauss–Weierstrass types. We continue with Chapter 11 which is about the approximation by interpolating operators induced by neural networks, a connection with computer science, a very detailed and extensive work covering all aspects of the topic. We finish with Chapter 12 about approximation theory and functional analysis on time scales, a very modern topic, detailing all the pros and cons of the approach.

The book's results are expected to find applications in many areas of pure and applied mathematics. As such this monograph is suitable for researchers, graduate students, and seminars of the above subjects, also to be in all science libraries.

The preparation of book took place during 2014–2015 in Memphis, TN, USA.

I would like to thank Professor Razvan Mezei, of Lenoir Rhyne University, for checking and reading the manuscript.

George A. Anastassiou
Department of Mathematical Sciences
University of Memphis
Memphis, TN 38152
USA
March 1, 2015

Contents

Chapter 1

Fractional Monotone Approximation

Let $f \in C^p([-1,1])$, $p \geq 0$ and let L be a linear left fractional differential operator such that $L(f) \geq 0$ throughout $[0,1]$. We can find a sequence of polynomials Q_n of degree $\leq n$ such that $L(Q_n) \geq 0$ over $[0,1]$, furthermore f is approximated uniformly by Q_n. The degree of this restricted approximations is given by an inequalities using the modulus of continuity of $f^{(p)}$.

This chapter follows [3].

1.1 Introduction

The topic of monotone approximation started in [6] has become a major trend in approximation theory. A typical problem in this subject is: given a positive integer k, approximate a given function whose kth derivative is ≥ 0 by polynomials having this property.

In [2] the authors replaced the kth derivative with a linear differential operator of order k. We mention this motivating result.

Theorem 1.1. *Let h, k, p be integers, $0 \leq h \leq k \leq p$ and let f be a real function, $f^{(p)}$ continuous in $[-1,1]$ with modulus of continuity $\omega_1(f^{(p)}, x)$ there. Let $a_j(x)$, $j = h, h+1, ..., k$ be real functions, defined and bounded on $[-1,1]$ and assume $a_h(x)$ is either $\geq$ some number $\alpha > 0$ or $\leq$ some number $\beta < 0$ throughout $[-1,1]$. Consider the operator*

$$L = \sum_{j=h}^{k} a_j(x) \left[\frac{d^j}{dx^j}\right] \tag{1.1}$$

and suppose, throughout $[-1,1]$,

$$L(f) \geq 0. \tag{1.2}$$

Then, for every integer $n \geq 1$, there is a real polynomial $Q_n(x)$ of degree $\leq n$ such that

$$L(Q_n) \geq 0 \text{ throughout } [-1,1] \tag{1.3}$$

and

$$\max_{-1\leq x\leq 1}|f(x)-Q_n(x)|\leq Cn^{k-p}\omega_1\left(f^{(p)},\frac{1}{n}\right), \tag{1.4}$$

where C is independent of n or f.

We need

Theorem (of Trigub see [7; 8]) Let $n\in\mathbb{N}$. Be given a real function g, with $g^{(p)}$ continuous in $[-1,1]$, there exists a real polynomial $q_n(x)$ of degree $\leq n$ such that $\max_{-1\leq x\leq 1}\left|g^{(j)}(x)-q_n^{(j)}(x)\right|\leq R_p n^{j-p}\omega_1\left(g^{(p)},\frac{1}{n}\right)$, $j=0,1,...,p$, where R_p is independent of n or g.

In this chapter we extend Theorem 1.1 to the fractional level. Now L is a linear left Caputo fractional differential operator. Here the monotonicity property is only true on the critical interval $[0,1]$. Quantitative uniform approximation remains true on all of $[-1,1]$.

To the best of our knowledge this is the first time fractional monotone Approximation Theory is introduced.

We need and make

Definition 1.1. ([4], p. 50) Let $\alpha>0$ and $\lceil\alpha\rceil=m$, ($\lceil\cdot\rceil$ ceiling of the number). Consider $f\in C^m([-1,1])$. We define the left Caputo fractional derivative of f of order α as follows:

$$\left(D_{*-1}^{\alpha}f\right)(x)=\frac{1}{\Gamma(m-\alpha)}\int_{-1}^{x}(x-t)^{m-\alpha-1}f^{(m)}(t)\,dt, \tag{1.5}$$

for any $x\in[-1,1]$, where Γ is the gamma function.

We set

$$D_{*-1}^{0}f(x)=f(x),$$
$$D_{*-1}^{m}f(x)=f^{(m)}(x),\ \ \forall\, x\in[-1,1]. \tag{1.6}$$

1.2 Main Result

We present

Theorem 1.2. *Let h,k,p be integers, $0\leq h\leq k\leq p$ and let f be a real function, $f^{(p)}$ continuous in $[-1,1]$ with modulus of continuity $\omega_1\left(f^{(p)},\delta\right)$, $\delta>0$, there. Let $\alpha_j(x)$, $j=h,h+1,...,k$ be real functions, defined and bounded on $[-1,1]$ and assume for $x\in[0,1]$ that $\alpha_h(x)$ is either $\geq$ some number $\alpha>0$ or $\leq$ some number $\beta<0$. Let the real numbers $\alpha_0=0<\alpha_1\leq 1<\alpha_2\leq 2<...<\alpha_p\leq p$. Here $D_{*-1}^{\alpha_j}f$ stands for the left Caputo fractional derivative of f of order α_j anchored at -1. Consider the linear left fractional differential operator*

$$L:=\sum_{j=h}^{k}\alpha_j(x)\left[D_{*-1}^{\alpha_j}\right] \tag{1.7}$$

and suppose, throughout $[0,1]$,

$$L(f) \geq 0. \tag{1.8}$$

Then, for any $n \in \mathbb{N}$, *there exists a real polynomial* $Q_n(x)$ *of degree* $\leq n$ *such that*

$$L(Q_n) \geq 0 \;\; \textit{throughout} \;\; [0,1], \tag{1.9}$$

and

$$\max_{-1 \leq x \leq 1} |f(x) - Q_n(x)| \leq C n^{k-p} \omega_1 \left(f^{(p)}, \frac{1}{n} \right), \tag{1.10}$$

where C *is independent of* n *or* f.

Proof. Let $n \in \mathbb{N}$. By the theorem of Trigub given a real function g, with $g^{(p)}$ continuous in $[-1,1]$, there exists a real polynomial $q_n(x)$ of degree $\leq n$ such that

$$\max_{-1 \leq x \leq 1} \left| g^{(j)}(x) - q_n^{(j)}(x) \right| \leq R_p n^{j-p} \omega_1 \left(g^{(p)}, \frac{1}{n} \right), \tag{1.11}$$

$j = 0, 1, ..., p$, where R_p is independent of n or g.

Here $h, k, p \in \mathbb{Z}_+$, $0 \leq h \leq k \leq p$.

Let $\alpha_j > 0$, $j = 1, ..., p$, such that $0 < \alpha_1 \leq 1 < \alpha_2 \leq 2 < \alpha_3 \leq 3... < ... < \alpha_p \leq p$. That is $\lceil \alpha_j \rceil = j$, $j = 1, ..., p$.

We consider the left Caputo fractional derivatives

$$\left(D_{*-1}^{\alpha_j} g\right)(x) = \frac{1}{\Gamma(j - \alpha_j)} \int_{-1}^{x} (x-t)^{j-\alpha_j-1} g^{(j)}(t)\, dt, \tag{1.12}$$

$$\left(D_{*-1}^{j} g\right)(x) = g^{(j)}(x),$$

and

$$\left(D_{*-1}^{\alpha_j} q_n\right)(x) = \frac{1}{\Gamma(j - \alpha_j)} \int_{-1}^{x} (x-t)^{j-\alpha_j-1} q_n^{(j)}(t)\, dt, \tag{1.13}$$

$$\left(D_{*-1}^{j} q_n\right)(x) = q_n^{(j)}(x); \; j = 1, ..., p,$$

where Γ is the gamma function

$$\Gamma(v) = \int_0^\infty e^{-t} t^{v-1} dt, \;\; v > 0. \tag{1.14}$$

We notice that

$$\begin{aligned} &\left| \left(D_{*-1}^{\alpha_j} g\right)(x) - \left(D_{*-1}^{\alpha_j} q_n\right)(x) \right| \\ &= \frac{1}{\Gamma(j-\alpha_j)} \left| \int_{-1}^{x} (x-t)^{j-\alpha_j-1} g^{(j)}(t)\, dt - \int_{-1}^{x} (x-t)^{j-\alpha_j-1} q_n^{(j)}(t)\, dt \right| \\ &= \frac{1}{\Gamma(j-\alpha_j)} \left| \int_{-1}^{x} (x-t)^{j-\alpha_j-1} \left(g^{(j)}(t) - q_n^{(j)}(t) \right) dt \right| \end{aligned} \tag{1.15}$$

$$\leq \frac{1}{\Gamma(j-\alpha_j)} \int_{-1}^{x} (x-t)^{j-\alpha_j-1} \left| g^{(j)}(t) - q_n^{(j)}(t) \right| dt \tag{1.16}$$

$$\overset{(1.11)}{\leq} \frac{1}{\Gamma(j-\alpha_j)} \left(\int_{-1}^{x} (x-t)^{j-\alpha_j-1} dt \right) R_p n^{j-p} \omega_1 \left(g^{(p)}, \frac{1}{n} \right)$$

$$= \frac{1}{\Gamma(j-\alpha_j)} \frac{(x+1)^{j-\alpha_j}}{(j-\alpha_j)} R_p n^{j-p} \omega_1 \left(g^{(p)}, \frac{1}{n} \right) \tag{1.17}$$

$$\leq \frac{2^{j-\alpha_j}}{\Gamma(j-\alpha_j+1)} R_p n^{j-p} \omega_1 \left(g^{(p)}, \frac{1}{n} \right).$$

We proved that for any $x \in [-1,1]$ we have

$$\left| \left(D_{*-1}^{\alpha_j} g \right)(x) - \left(D_{*-1}^{\alpha_j} q_n \right)(x) \right| \leq \frac{2^{j-\alpha_j}}{\Gamma(j-\alpha_j+1)} R_p n^{j-p} \omega_1 \left(g^{(p)}, \frac{1}{n} \right). \tag{1.18}$$

Hence it holds

$$\max_{-1 \leq x \leq 1} \left| \left(D_{*-1}^{\alpha_j} g \right)(x) - \left(D_{*-1}^{\alpha_j} q_n \right)(x) \right| \leq \frac{2^{j-\alpha_j}}{\Gamma(j-\alpha_j+1)} R_p n^{j-p} \omega_1 \left(g^{(p)}, \frac{1}{n} \right), \tag{1.19}$$

$j = 0, 1, ..., p$.

Above we set $D_{*-1}^{0} g(x) = g(x)$, $D_{*-1}^{0} q_n(x) = q_n(x)$, $\forall\, x \in [-1,1]$, and $\alpha_0 = 0$, i.e. $\lceil \alpha_0 \rceil = 0$.

Put

$$s_j \equiv \sup_{-1 \leq x \leq 1} \left| \alpha_h^{-1}(x) \alpha_j(x) \right|, \quad j = h, ..., k, \tag{1.20}$$

and

$$\eta_n := R_p \omega_1 \left(f^{(p)}, \frac{1}{n} \right) \left(\sum_{j=h}^{k} s_j \frac{2^{j-\alpha_j}}{\Gamma(j-\alpha_j+1)} n^{j-p} \right). \tag{1.21}$$

I. Suppose, throughout $[0,1]$, $\alpha_h(x) \geq \alpha > 0$. Let $Q_n(x)$, $x \in [-1,1]$, be a real polynomial of degree $\leq n$ so that

$$\max_{-1 \leq x \leq 1} \left| D_{*-1}^{\alpha_j} \left(f(x) + \eta_n (h!)^{-1} x^h \right) - \left(D_{*-1}^{\alpha_j} Q_n \right)(x) \right|$$

$$\overset{(1.19)}{\leq} \frac{2^{j-\alpha_j}}{\Gamma(j-\alpha_j+1)} R_p n^{j-p} \omega_1 \left(f^{(p)}, \frac{1}{n} \right), \tag{1.22}$$

$j = 0, 1, ..., p$.

In particular ($j = 0$) holds

$$\max_{-1 \leq x \leq 1} \left| \left(f(x) + \eta_n (h!)^{-1} x^h \right) - Q_n(x) \right| \leq R_p n^{-p} \omega_1 \left(f^{(p)}, \frac{1}{n} \right), \tag{1.23}$$

and

$$\max_{-1 \leq x \leq 1} |f(x) - Q_n(x)| \leq \eta_n (h!)^{-1} + R_p n^{-p} \omega_1 \left(f^{(p)}, \frac{1}{n} \right)$$

$$= (h!)^{-1} R_p \omega_1 \left(f^{(p)}, \frac{1}{n} \right) \left(\sum_{j=h}^{k} s_j \frac{2^{j-\alpha_j}}{\Gamma(j - \alpha_j + 1)} n^{j-p} \right)$$

$$+ R_p n^{-p} \omega_1 \left(f^{(p)}, \frac{1}{n} \right) \tag{1.24}$$

$$\leq R_p \omega_1 \left(f^{(p)}, \frac{1}{n} \right) n^{k-p} \left(1 + (h!)^{-1} \sum_{j=h}^{k} s_j \frac{2^{j-\alpha_j}}{\Gamma(j - \alpha_j + 1)} \right). \tag{1.25}$$

That is

$$\max_{-1 \leq x \leq 1} |f(x) - Q_n(x)|$$

$$\leq R_p \left(1 + (h!)^{-1} \sum_{j=h}^{k} s_j \frac{2^{j-\alpha_j}}{\Gamma(j - \alpha_j + 1)} \right) n^{k-p} \omega_1 \left(f^{(p)}, \frac{1}{n} \right), \tag{1.26}$$

proving (1.10).

Here

$$L = \sum_{j=h}^{k} \alpha_j(x) \left[D_{*-1}^{\alpha_j} \right],$$

and suppose, throughout $[0, 1]$, $Lf \geq 0$.

So over $0 \leq x \leq 1$, using (1.12) to compute $D_{*-1}^{\alpha_j} x^h$, (1.21) for η_n and (1.22), we get

$$\alpha_h^{-1}(x) L(Q_n(x)) = \alpha_h^{-1}(x) L(f(x)) + \eta_n \frac{(x+1)^{h-\alpha_h}}{\Gamma(h - \alpha_h + 1)} \tag{1.27}$$

$$+ \sum_{j=h}^{k} \alpha_h^{-1}(x) \alpha_j(x) \left[D_{*-1}^{\alpha_j} Q_n(x) - D_{*-1}^{\alpha_j} f(x) - \frac{\eta_n}{h!} D_{*-1}^{\alpha_j} x^h \right]$$

$$\geq \eta_n \frac{(x+1)^{h-\alpha_h}}{\Gamma(h - \alpha_h + 1)} - \left(\sum_{j=h}^{k} s_j \frac{2^{j-\alpha_j}}{\Gamma(j - \alpha_j + 1)} n^{j-p} \right) R_p \omega_1 \left(f^{(p)}, \frac{1}{n} \right)$$

$$= \eta_n \frac{(x+1)^{h-\alpha_h}}{\Gamma(h - \alpha_h + 1)} - \eta_n = \eta_n \left[\frac{(x+1)^{h-\alpha_h}}{\Gamma(h - \alpha_h + 1)} - 1 \right] \tag{1.28}$$

$$= \eta_n \left[\frac{(x+1)^{h-\alpha_h} - \Gamma(h-\alpha_h+1)}{\Gamma(h-\alpha_h+1)} \right] \geq \eta_n \left[\frac{1-\Gamma(h-\alpha_h+1)}{\Gamma(h-\alpha_h+1)} \right] \geq 0. \quad (1.29)$$

Explanation: We know $\Gamma(1) = 1$, $\Gamma(2) = 1$, and Γ is convex and positive on $(0, \infty)$. Here $0 \leq h - \alpha_h < 1$ and $1 \leq h - \alpha_h + 1 < 2$. Thus $\Gamma(h-\alpha_h+1) \leq 1$ and

$$1 - \Gamma(h-\alpha_h+1) \geq 0. \quad (1.30)$$

Hence

$$L(Q_n(x)) \geq 0, x \in [0,1]. \quad (1.31)$$

II. Suppose, throughout $[0,1]$, $\alpha_h(x) \leq \beta < 0$. In this case let $Q_n(x)$, $x \in [-1,1]$, be a real polynomial of degree $\leq n$ such that

$$\max_{-1\leq x\leq 1} \left| D_{*-1}^{\alpha_j} \left(f(x) - \eta_n (h!)^{-1} x^h \right) - \left(D_{*-1}^{\alpha_j} Q_n \right)(x) \right| \quad (1.32)$$

$$\leq \frac{2^{j-\alpha_j}}{\Gamma(j-\alpha_j+1)} R_p n^{j-p} \omega_1 \left(f^{(p)}, \frac{1}{n} \right),$$

$j = 0, 1, ..., p$.

In particular holds ($j = 0$)

$$\max_{-1\leq x\leq 1} \left| \left(f(x) - \eta_n (h!)^{-1} x^h \right) - Q_n(x) \right| \leq R_p n^{-p} \omega_1 \left(f^{(p)}, \frac{1}{n} \right), \quad (1.33)$$

and

$$\max_{-1\leq x\leq 1} |f(x) - Q_n(x)| \leq \eta_n (h!)^{-1} + R_p n^{-p} \omega_1 \left(f^{(p)}, \frac{1}{n} \right)$$

$$\overset{\text{(as before)}}{\leq} R_p \omega_1 \left(f^{(p)}, \frac{1}{n} \right) n^{k-p} \left(1 + (h!)^{-1} \sum_{j=h}^{k} s_j \frac{2^{j-\alpha_j}}{\Gamma(j-\alpha_j+1)} \right). \quad (1.34)$$

That is

$$\max_{-1\leq x\leq 1} |f(x) - Q_n(x)|$$

$$\leq R_p \left(1 + (h!)^{-1} \sum_{j=h}^{k} s_j \frac{2^{j-\alpha_j}}{\Gamma(j-\alpha_j+1)} \right) n^{k-p} \omega_1 \left(f^{(p)}, \frac{1}{n} \right), \quad (1.35)$$

reproving (1.10).

Again suppose, throughout $[0,1]$, $Lf \geq 0$.

Also if $0 \leq x \leq 1$, then

$$\alpha_h^{-1}(x) L(Q_n(x)) = \alpha_h^{-1}(x) L(f(x)) - \eta_n \frac{(x+1)^{h-\alpha_h}}{\Gamma(h-\alpha_h+1)} \quad (1.36)$$

$$+ \sum_{j=h}^{k} \alpha_h^{-1}(x) \alpha_j(x) \left[D_{*-1}^{\alpha_j} Q_n(x) - D_{*-1}^{\alpha_j} f(x) + \frac{\eta_n}{h!} \left(D_{*-1}^{\alpha_j} x^h \right) \right]$$

$$\overset{(1.32)}{\leq} -\eta_n \frac{(x+1)^{h-\alpha_h}}{\Gamma(h-\alpha_h+1)} + \left(\sum_{j=h}^{k} s_j \frac{2^{j-\alpha_j}}{\Gamma(j-\alpha_j+1)} n^{j-p} \right) R_p \omega_1 \left(f^{(p)}, \frac{1}{n} \right)$$

$$= \eta_n - \eta_n \frac{(x+1)^{h-\alpha_h}}{\Gamma(h-\alpha_h+1)} = \eta_n \left(1 - \frac{(x+1)^{h-\alpha_h}}{\Gamma(h-\alpha_h+1)} \right)$$

$$= \eta_n \left(\frac{\Gamma(h-\alpha_h+1) - (x+1)^{h-\alpha_h}}{\Gamma(h-\alpha_h+1)} \right) \leq \eta_n \left(\frac{1-(x+1)^{h-\alpha_h}}{\Gamma(h-\alpha_h+1)} \right) \leq 0, \tag{1.37}$$

and hence again

$$L(Q_n(x)) \geq 0, \quad \forall\, x \in [0,1]. \tag{1.38}$$

□

Remark 1.1. Based on [1], here we have that $D_{*-1}^{\alpha_j} f$ are continuous functions, $j = 0, 1, ..., p$. Suppose that $\alpha_h(x), ..., \alpha_k(x)$ are continuous functions in $[-1, 1]$, and $L(f) \geq 0$ on $[0, 1]$ is replaced by $L(f) > 0$ on $[0, 1]$. Disregard the assumption made in the Theorem 1.2 on $\alpha_h(x)$. For $n \in \mathbb{N}$, let $Q_n(x)$ be $q_n(x)$ of (1.19) for $g = f$. Then $Q_n(x)$ converges to f at the Jackson rate [5, p. 18, Theorem VIII] and at the same time, since $L(Q_n)$ converges uniformly to $L(f)$ on $[-1, 1]$, $L(Q_n) > 0$ on $[0, 1]$ for all n sufficiently large.

Bibliography

1. G.A. Anastassiou, *Fractional Korovkin theory*, Chaos, Solitons Fractals 42 (2009), 2080-2094.
2. G.A. Anastassiou, O. Shisha, *Monotone approximation with linear differential operators*, J. Approx. Theory 44 (1985), 391-393.
3. G.A. Anastassiou, *Fractional monotone approximation theory*, Indian Journal of Mathematics, to appear 2015.
4. K. Diethelm, *The Analysis of Fractional Differential Equations*, Lecture Notes in Mathematics, Vol. 2004, 1st edition, Springer, New York, Heidelberg, 2010.
5. D. Jackson, *The Theory of Approximation*, Amer. Math. Soc. Colloq., Vol. XI, New York, 1930.
6. O. Shisha, *Monotone approximation*, Pacific J. Math. 15 (1965), 667-671.
7. S.A. Teljakovskii, *Two theorems on the approximation of functions by algebraic polynomials*, Mat. Sb. 70 (112) (1966), 252-265 [Russian]; Amer. Math. Soc. Trans. 77 (2) (1968), 163-178.
8. R.M. Trigub, *Approximation of functions by polynomials with integer coefficients*, Izv. Akad. Nauk SSSR Ser. Mat. 26 (1962), 261-280 [Russian].

Chapter 2

Right Fractional Monotone Approximation Theory

Let $f \in C^p([-1,1])$, $p \geq 0$ and let L be a linear right fractional differential operator such that $L(f) \geq 0$ throughout $[-1,0]$. We can find a sequence of polynomials Q_n of degree $\leq n$ such that $L(Q_n) \geq 0$ over $[-1,0]$, furthermore f is approximated uniformly by Q_n. The degree of this restricted approximations is given by an inequalities using the modulus of continuity of $f^{(p)}$.

2.1 Introduction

The topic of monotone approximation started in [5] has become a major trend in approximation theory. A typical problem in this subject is: given a positive integer k, approximate a given function whose kth derivative is ≥ 0 by polynomials having this property.

In [2] the authors replaced the kth derivative with a linear differential operator of order k. We mention this motivating result.

Theorem 2.1. *Let h, k, p be integers, $0 \leq h \leq k \leq p$ and let f be a real function, $f^{(p)}$ continuous in $[-1,1]$ with modulus of continuity $\omega_1(f^{(p)}, x)$ there. Let $a_j(x)$, $j = h, h+1, ..., k$ be real functions, defined and bounded on $[-1,1]$ and assume $a_h(x)$ is either $\geq$ some number $\alpha > 0$ or $\leq$ some number $\beta < 0$ throughout $[-1,1]$. Consider the operator*

$$L = \sum_{j=h}^{k} a_j(x) \left[\frac{d^j}{dx^j} \right] \tag{2.1}$$

and suppose, throughout $[-1,1]$,

$$L(f) \geq 0. \tag{2.2}$$

Then, for every integer $n \geq 1$, there is a real polynomial $Q_n(x)$ of degree $\leq n$ such that

$$L(Q_n) \geq 0 \text{ throughout } [-1,1] \tag{2.3}$$

and

$$\max_{-1\leq x\leq 1}\left|f(x)-Q_n(x)\right|\leq Cn^{k-p}\omega_1\left(f^{(p)},\frac{1}{n}\right), \tag{2.4}$$

where C is independent of n or f.

In this chapter we extend Theorem 2.1 to the right fractional level. Now L is a linear right Caputo fractional differential operator. Here the monotonicity property is only true on the critical interval $[-1,0]$. Quantitative uniform approximation remains true on all of $[-1,1]$.

To the best of our knowledge this is the first time right fractional monotone Approximation is introduced.

We need and make

Definition 2.1. ([3]) Let $\alpha > 0$ and $\lceil\alpha\rceil = m$, ($\lceil\cdot\rceil$ ceiling of the number). Consider $f \in C^m([-1,1])$. We define the right Caputo fractional derivative of f of order α as follows:

$$\left(D_{1-}^{\alpha}f\right)(x)=\frac{(-1)^m}{\Gamma(m-\alpha)}\int_x^1(t-x)^{m-\alpha-1}f^{(m)}(t)\,dt, \tag{2.5}$$

for any $x\in[-1,1]$, where Γ is the gamma function.

We set

$$D_{1-}^{0}f(x)=f(x),$$

$$D_{1-}^{m}f(x)=(-1)^m f^{(m)}(x),\ \ \forall\, x\in[-1,1]. \tag{2.6}$$

2.2 Main Result

We present

Theorem 2.2. *Let h, k, p be integers, h is even, $0\leq h\leq k\leq p$ and let f be a real function, $f^{(p)}$ continuous in $[-1,1]$ with modulus of continuity $\omega_1\left(f^{(p)},\delta\right)$, $\delta>0$, there. Let $\alpha_j(x)$, $j=h,h+1,...,k$ be real functions, defined and bounded on $[-1,1]$ and assume for $x\in[-1,0]$ that $\alpha_h(x)$ is either $\geq$ some number $\alpha>0$ or $\leq$ some number $\beta<0$. Let the real numbers $\alpha_0=0<\alpha_1<1<\alpha_2<2<...<\alpha_p<p$. Here $D_{1-}^{\alpha_j}f$ stands for the right Caputo fractional derivative of f of order α_j anchored at 1. Consider the linear right fractional differential operator*

$$L:=\sum_{j=h}^{k}\alpha_j(x)\left[D_{1-}^{\alpha_j}\right] \tag{2.7}$$

and suppose, throughout $[-1,0]$,

$$L(f)\geq 0. \tag{2.8}$$

Then, for any $n \in \mathbb{N}$, there exists a real polynomial $Q_n(x)$ of degree $\leq n$ such that

$$L(Q_n) \geq 0 \;\; throughout \;\; [-1,0], \tag{2.9}$$

and

$$\max_{-1 \leq x \leq 1} |f(x) - Q_n(x)| \leq C n^{k-p} \omega_1\left(f^{(p)}, \frac{1}{n}\right), \tag{2.10}$$

where C is independent of n or f.

Proof. Let $n \in \mathbb{N}$. By a theorem of Trigub [6; 7], given a real function g, with $g^{(p)}$ continuous in $[-1,1]$, there exists a real polynomial $q_n(x)$ of degree $\leq n$ such that

$$\max_{-1 \leq x \leq 1} \left| g^{(j)}(x) - q_n^{(j)}(x) \right| \leq R_p n^{j-p} \omega_1\left(g^{(p)}, \frac{1}{n}\right), \tag{2.11}$$

$j = 0, 1, ..., p$, where R_p is independent of n or g.

Here $h, k, p \in \mathbb{Z}_+$, $0 \leq h \leq k \leq p$.

Let $\alpha_j > 0$, $j = 1, ..., p$, such that $0 < \alpha_1 < 1 < \alpha_2 < 2 < \alpha_3 < 3... < ... < \alpha_p < p$. That is $\lceil \alpha_j \rceil = j$, $j = 1, ..., p$.

We consider the right Caputo fractional derivatives

$$\left(D_{1-}^{\alpha_j} g\right)(x) = \frac{(-1)^j}{\Gamma(j-\alpha_j)} \int_x^1 (t-x)^{j-\alpha_j-1} g^{(j)}(t)\, dt, \tag{2.12}$$

$$\left(D_{1-}^{j} g\right)(x) = (-1)^j g^{(j)}(x),$$

and

$$\left(D_{1-}^{\alpha_j} q_n\right)(x) = \frac{(-1)^j}{\Gamma(j-\alpha_j)} \int_x^1 (t-x)^{j-\alpha_j-1} q_n^{(j)}(t)\, dt, \tag{2.13}$$

$$\left(D_{1-}^{j} q_n\right)(x) = (-1)^j q_n^{(j)}(x); \; j = 1, ..., p,$$

where Γ is the gamma function

$$\Gamma(v) = \int_0^\infty e^{-t} t^{v-1} dt, \quad v > 0. \tag{2.14}$$

We notice that

$$\left| \left(D_{1-}^{\alpha_j} g\right)(x) - \left(D_{1-}^{\alpha_j} q_n\right)(x) \right|$$

$$= \frac{1}{\Gamma(j-\alpha_j)} \left| \int_x^1 (t-x)^{j-\alpha_j-1} g^{(j)}(t)\, dt - \int_x^1 (t-x)^{j-\alpha_j-1} q_n^{(j)}(t)\, dt \right| \tag{2.15}$$

$$= \frac{1}{\Gamma(j-\alpha_j)} \left| \int_x^1 (t-x)^{j-\alpha_j-1} \left(g^{(j)}(t) - q_n^{(j)}(t) \right) dt \right|$$

$$\leq \frac{1}{\Gamma(j-\alpha_j)} \int_x^1 (t-x)^{j-\alpha_j-1} \left| g^{(j)}(t) - q_n^{(j)}(t) \right| dt \tag{2.16}$$

$$\overset{(2.11)}{\leq} \frac{1}{\Gamma(j-\alpha_j)} \left(\int_x^1 (t-x)^{j-\alpha_j-1} dt \right) R_p n^{j-p} \omega_1 \left(g^{(p)}, \frac{1}{n} \right)$$

$$= \frac{1}{\Gamma(j-\alpha_j)} \frac{(1-x)^{j-\alpha_j}}{(j-\alpha_j)} R_p n^{j-p} \omega_1 \left(g^{(p)}, \frac{1}{n} \right) \tag{2.17}$$

$$\leq \frac{2^{j-\alpha_j}}{\Gamma(j-\alpha_j+1)} R_p n^{j-p} \omega_1 \left(g^{(p)}, \frac{1}{n} \right).$$

We proved that for any $x \in [-1,1]$ we have

$$\left| \left(D_{1-}^{\alpha_j} g \right)(x) - \left(D_{1-}^{\alpha_j} q_n \right)(x) \right| \leq \frac{2^{j-\alpha_j}}{\Gamma(j-\alpha_j+1)} R_p n^{j-p} \omega_1 \left(g^{(p)}, \frac{1}{n} \right). \tag{2.18}$$

Hence it holds

$$\max_{-1 \leq x \leq 1} \left| \left(D_{1-}^{\alpha_j} g \right)(x) - \left(D_{1-}^{\alpha_j} q_n \right)(x) \right| \leq \frac{2^{j-\alpha_j}}{\Gamma(j-\alpha_j+1)} R_p n^{j-p} \omega_1 \left(g^{(p)}, \frac{1}{n} \right), \tag{2.19}$$

$j = 0, 1, ..., p$.

Above we set $D_{1-}^0 g(x) = g(x)$, $D_{1-}^0 q_n(x) = q_n(x)$, $\forall\ x \in [-1,1]$, and $\alpha_0 = 0$, i.e. $\lceil \alpha_0 \rceil = 0$.

Put

$$s_j \equiv \sup_{-1 \leq x \leq 1} \left| \alpha_h^{-1}(x) \alpha_j(x) \right|, \ \ j = h, ..., k, \tag{2.20}$$

and

$$\eta_n := R_p \omega_1 \left(f^{(p)}, \frac{1}{n} \right) \left(\sum_{j=h}^{k} s_j \frac{2^{j-\alpha_j}}{\Gamma(j-\alpha_j+1)} n^{j-p} \right). \tag{2.21}$$

I. Suppose, throughout $[-1,0]$, $\alpha_h(x) \geq \alpha > 0$. Let $Q_n(x)$, $x \in [-1,1]$, be a real polynomial of degree $\leq n$ so that

$$\max_{-1 \leq x \leq 1} \left| D_{1-}^{\alpha_j} \left(f(x) + \eta_n (h!)^{-1} x^h \right) - \left(D_{1-}^{\alpha_j} Q_n \right)(x) \right|$$

$$\overset{(2.19)}{\leq} \frac{2^{j-\alpha_j}}{\Gamma(j-\alpha_j+1)} R_p n^{j-p} \omega_1 \left(f^{(p)}, \frac{1}{n} \right), \tag{2.22}$$

$j = 0, 1, ..., p$.

In particular ($j = 0$) holds

$$\max_{-1 \leq x \leq 1} \left| \left(f(x) + \eta_n (h!)^{-1} x^h \right) - Q_n(x) \right| \leq R_p n^{-p} \omega_1 \left(f^{(p)}, \frac{1}{n} \right), \tag{2.23}$$

and

$$\max_{-1 \leq x \leq 1} \left| f(x) - Q_n(x) \right| \leq \eta_n (h!)^{-1} + R_p n^{-p} \omega_1 \left(f^{(p)}, \frac{1}{n} \right)$$

$$= (h!)^{-1} R_p \omega_1 \left(f^{(p)}, \frac{1}{n} \right) \left(\sum_{j=h}^{k} s_j \frac{2^{j-\alpha_j}}{\Gamma(j-\alpha_j+1)} n^{j-p} \right)$$

$$+ R_p n^{-p} \omega_1 \left(f^{(p)}, \frac{1}{n} \right) \tag{2.24}$$

$$\leq R_p \omega_1 \left(f^{(p)}, \frac{1}{n} \right) n^{k-p} \left(1 + (h!)^{-1} \sum_{j=h}^{k} s_j \frac{2^{j-\alpha_j}}{\Gamma(j-\alpha_j+1)} \right). \tag{2.25}$$

That is

$$\max_{-1 \leq x \leq 1} |f(x) - Q_n(x)|$$

$$\leq R_p \left(1 + (h!)^{-1} \sum_{j=h}^{k} s_j \frac{2^{j-\alpha_j}}{\Gamma(j-\alpha_j+1)} \right) n^{k-p} \omega_1 \left(f^{(p)}, \frac{1}{n} \right), \tag{2.26}$$

proving (2.10).

Here

$$L = \sum_{j=h}^{k} \alpha_j(x) \left[D_{1-}^{\alpha_j} \right],$$

and suppose, throughout $[-1, 0]$, $Lf \geq 0$.

So over $-1 \leq x \leq 0$, we get

$$\alpha_h^{-1}(x) L(Q_n(x)) = \alpha_h^{-1}(x) L(f(x)) + \eta_n \frac{(1-x)^{h-\alpha_h}}{\Gamma(h-\alpha_h+1)} \tag{2.27}$$

$$+ \sum_{j=h}^{k} \alpha_h^{-1}(x) \alpha_j(x) \left[D_{1-}^{\alpha_j} Q_n(x) - D_{1-}^{\alpha_j} f(x) - \frac{\eta_n}{h!} D_{1-}^{\alpha_j} x^h \right]$$

$$\overset{(2.22)}{\geq} \eta_n \frac{(1-x)^{h-\alpha_h}}{\Gamma(h-\alpha_h+1)} - \left(\sum_{j=h}^{k} s_j \frac{2^{j-\alpha_j}}{\Gamma(j-\alpha_j+1)} n^{j-p} \right) R_p \omega_1 \left(f^{(p)}, \frac{1}{n} \right)$$

$$= \eta_n \frac{(1-x)^{h-\alpha_h}}{\Gamma(h-\alpha_h+1)} - \eta_n = \eta_n \left[\frac{(1-x)^{h-\alpha_h}}{\Gamma(h-\alpha_h+1)} - 1 \right] \tag{2.28}$$

$$= \eta_n \left[\frac{(1-x)^{h-\alpha_h} - \Gamma(h-\alpha_h+1)}{\Gamma(h-\alpha_h+1)} \right] \geq \eta_n \left[\frac{1 - \Gamma(h-\alpha_h+1)}{\Gamma(h-\alpha_h+1)} \right] \geq 0. \tag{2.29}$$

Explanation: We know $\Gamma(1) = 1$, $\Gamma(2) = 1$, and Γ is convex and positive on $(0, \infty)$. Here $0 \leq h - \alpha_h < 1$ and $1 \leq h - \alpha_h + 1 < 2$. Thus $\Gamma(h - \alpha_h + 1) \leq 1$ and

$$1 - \Gamma(h - \alpha_h + 1) \geq 0. \tag{2.30}$$

Hence

$$L\left(Q_n(x)\right) \geq 0, x \in [-1,0]. \tag{2.31}$$

II. Suppose, throughout $[-1,0]$, $\alpha_h(x) \leq \beta < 0$. In this case let $Q_n(x)$, $x \in [-1,1]$, be a real polynomial of degree $\leq n$ such that

$$\max_{-1\leq x\leq 1}\left|D_{1-}^{\alpha_j}\left(f(x) - \eta_n (h!)^{-1} x^h\right) - \left(D_{1-}^{\alpha_j} Q_n\right)(x)\right| \tag{2.32}$$
$$\leq \frac{2^{j-\alpha_j}}{\Gamma(j-\alpha_j+1)} R_p n^{j-p} \omega_1\left(f^{(p)}, \frac{1}{n}\right),$$

$j = 0, 1, ..., p$.

In particular holds ($j = 0$)

$$\max_{-1\leq x\leq 1}\left|\left(f(x) - \eta_n (h!)^{-1} x^h\right) - Q_n(x)\right| \leq R_p n^{-p} \omega_1\left(f^{(p)}, \frac{1}{n}\right), \tag{2.33}$$

and

$$\max_{-1\leq x\leq 1}\left|f(x) - Q_n(x)\right| \leq \eta_n (h!)^{-1} + R_p n^{-p} \omega_1\left(f^{(p)}, \frac{1}{n}\right)$$

$$\overset{\text{(as before)}}{\leq} R_p \omega_1\left(f^{(p)}, \frac{1}{n}\right) n^{k-p}\left(1 + (h!)^{-1}\sum_{j=h}^{k} s_j \frac{2^{j-\alpha_j}}{\Gamma(j-\alpha_j+1)}\right). \tag{2.34}$$

That is

$$\max_{-1\leq x\leq 1}\left|f(x) - Q_n(x)\right|$$

$$\leq R_p\left(1 + (h!)^{-1}\sum_{j=h}^{k} s_j \frac{2^{j-\alpha_j}}{\Gamma(j-\alpha_j+1)}\right) n^{k-p} \omega_1\left(f^{(p)}, \frac{1}{n}\right), \tag{2.35}$$

reproving (2.10).

Again suppose, throughout $[-1,0]$, $Lf \geq 0$.

Also if $-1 \leq x \leq 0$, then

$$\alpha_h^{-1}(x) L\left(Q_n(x)\right) = \alpha_h^{-1}(x) L\left(f(x)\right) - \eta_n \frac{(1-x)^{h-\alpha_h}}{\Gamma(h-\alpha_h+1)} \tag{2.36}$$

$$+\sum_{j=h}^{k} \alpha_h^{-1}(x) \alpha_j(x)\left[D_{1-}^{\alpha_j} Q_n(x) - D_{1-}^{\alpha_j} f(x) + \frac{\eta_n}{h!}\left(D_{1-}^{\alpha_j} x^h\right)\right]$$

$$\overset{(2.32)}{\leq} -\eta_n \frac{(1-x)^{h-\alpha_h}}{\Gamma(h-\alpha_h+1)} + \left(\sum_{j=h}^{k} s_j \frac{2^{j-\alpha_j}}{\Gamma(j-\alpha_j+1)} n^{j-p}\right) R_p \omega_1\left(f^{(p)}, \frac{1}{n}\right)$$

$$= \eta_n - \eta_n \frac{(1-x)^{h-\alpha_h}}{\Gamma(h-\alpha_h+1)} = \eta_n\left(1 - \frac{(1-x)^{h-\alpha_h}}{\Gamma(h-\alpha_h+1)}\right)$$

$$= \eta_n\left(\frac{\Gamma(h-\alpha_h+1) - (1-x)^{h-\alpha_h}}{\Gamma(h-\alpha_h+1)}\right) \leq \eta_n\left(\frac{1-(1-x)^{h-\alpha_h}}{\Gamma(h-\alpha_h+1)}\right) \leq 0, \tag{2.37}$$

and hence again

$$L\left(Q_n(x)\right) \geq 0, \quad \forall\, x \in [-1,0]. \tag{2.38}$$

□

Remark 2.1. Based on [1], here we have that $D_{1-}^{\alpha_j} f$ are continuous functions, $j = 0, 1, ..., p$. Suppose that $\alpha_h(x), ..., \alpha_k(x)$ are continuous functions in $[-1, 1]$, and $L(f) \geq 0$ on $[-1, 0]$ is replaced by $L(f) > 0$ on $[-1, 0]$. Disregard the assumption made in the Theorem 2.2 on $\alpha_h(x)$. For $n \in \mathbb{N}$, let $Q_n(x)$ be $q_n(x)$ of (2.19) for $g = f$. Then $Q_n(x)$ converges to f at the Jackson rate [4, p. 18, Theorem VIII] and at the same time, since $L(Q_n)$ converges uniformly to $L(f)$ on $[-1, 1]$, $L(Q_n) > 0$ on $[-1, 0]$ for all n sufficiently large.

Bibliography

1. G.A. Anastassiou, *Fractional Korovkin theory*, Chaos, Solitons Fractals 42 (2009), 2080-2094.
2. G.A. Anastassiou, O. Shisha, *Monotone approximation with linear differential operators*, J. Approx. Theory 44 (1985), 391-393.
3. A.M.A. El-Sayed, M. Gaber, *On the finite Caputo and finite Riesz derivatives*, Electron. J. Theoret. Phys. 3(12) (2006), 81-95.
4. D. Jackson, *The Theory of Approximation*, Amer. Math. Soc. Colloq., Vol. XI, New York, 1930.
5. O. Shisha, *Monotone approximation*, Pacific J. Math. 15 (1965), 667-671.
6. S.A. Teljakovskii, *Two theorems on the approximation of functions by algebraic polynomials*, Mat. Sb. 70 (112) (1966), 252-265 [Russian]; Amer. Math. Soc. Trans. 77 (2) (1968), 163-178.
7. R.M. Trigub, *Approximation of functions by polynomials with integer coefficients*, Izv. Akad. Nauk SSSR Ser. Mat. 26 (1962), 261-280 [Russian].

Chapter 3

Univariate Left Fractional Polynomial High Order Monotone Approximation Theory

Let $f \in C^r([-1,1])$, $r \geq 0$ and let L^* be a linear left fractional differential operator such that $L^*(f) \geq 0$ throughout $[0,1]$. We can find a sequence of polynomials Q_n of degree $\leq n$ such that $L^*(Q_n) \geq 0$ over $[0,1]$, furthermore f is approximated left fractionally and simultaneously by Q_n on $[-1,1]$. The degree of these restricted approximations is given via inequalities using a higher order modulus of smoothness for $f^{(r)}$. It follows [4].

3.1 Introduction

The topic of monotone approximation started in [7] has become a major trend in approximation theory. A typical problem in this subject is: given a positive integer k, approximate a given function whose kth derivative is ≥ 0 by polynomials having this property.

In [3] the authors replaced the kth derivative with a linear differential operator of order k. We mention this motivating result.

Theorem 3.1. *Let h, k, p be integers, $0 \leq h \leq k \leq p$ and let f be a real function, $f^{(p)}$ continuous in $[-1,1]$ with modulus of continuity $\omega_1(f^{(p)}, x)$ there. Let $a_j(x)$, $j = h, h+1, ..., k$ be real functions, defined and bounded on $[-1,1]$ and assume $a_h(x)$ is either $\geq$ some number $\alpha > 0$ or $\leq$ some number $\beta < 0$ throughout $[-1,1]$. Consider the operator*

$$L = \sum_{j=h}^{k} a_j(x) \left[\frac{d^j}{dx^j}\right]$$

and suppose, throughout $[-1,1]$,

$$L(f) \geq 0. \tag{3.1}$$

Then, for every integer $n \geq 1$, there is a real polynomial $Q_n(x)$ of degree $\leq n$ such that

$$L(Q_n) \geq 0 \text{ throughout } [-1,1]$$

and

$$\max_{-1\leq x\leq 1}|f(x)-Q_n(x)|\leq Cn^{k-p}\omega_1\left(f^{(p)},\frac{1}{n}\right),$$

where C is independent of n or f.

We use also the notation $I=[-1,1]$.

We would like to mention

Theorem 3.2. *(Gonska and Hinnemann [5]). Let $r\geq 0$ and $s\geq 1$. Then there exists a sequence $Q_n=Q_n^{(r,s)}$ of linear polynomial operators mapping $C^r(I)$ into P_n (space of polynomials of degree $\leq n$), such that for all $f\in C^r(I)$, all $|x|\leq 1$ and all $n\geq\max(4(r+1),r+s)$ we have*

$$\left|f^{(k)}(x)-(Q_nf)^{(k)}(x)\right|\leq M_{r,s}(\Delta_n(x))^{r-k}\omega_s\left(f^{(r)},\Delta_n(x)\right),\quad 0\leq k\leq r,\qquad (3.2)$$

where $\Delta_n(x)=\frac{\sqrt{1-x^2}}{n}+\frac{1}{n^2}$, and $M_{r,s}$ is a constant independent of f, x, and n. Above ω_s is the usual modulus of smoothness of order s with respect to the supremum norm.

Theorem 3.2 implies the useful

Corollary 3.1. *([2]) Let $r\geq 0$ and $s\geq 1$. Then there exists a sequence $Q_n=Q_n^{(r,s)}$ of linear polynomial operators mapping $C^r(I)$ into P_n, such that for all $f\in C^r(I)$ and all $n\geq\max(4(r+1),r+s)$ we have*

$$\left\|f^{(k)}-(Q_nf)^{(k)}\right\|_\infty\leq\frac{C_{r,s}}{n^{r-k}}\omega_s\left(f^{(r)},\frac{1}{n}\right),\quad k=0,1,...,r,\qquad (3.3)$$

where $C_{r,s}$ is a constant independent of f and n.

In [2] we proved the motivational

Theorem 3.3. *Let h,v,r be integers, $0\leq h\leq v\leq r$ and let $f\in C^r(I)$, with $f^{(r)}$ having modulus of smoothness $\omega_s\left(f^{(r)},\delta\right)$ there, $s\geq 1$. Let $\alpha_j(x)$, $j=h,h+1,...,v$ be real functions, defined and bounded on I and suppose α_h is either $\geq\alpha>0$ or $\leq\beta<0$ throughout I. Take the operator*

$$L=\sum_{j=h}^{v}\alpha_j(x)\left[\frac{d^j}{dx^j}\right]\qquad (3.4)$$

and assume, throughout I,

$$L(f)\geq 0.\qquad (3.5)$$

Then for every integer $n\geq\max(4(r+1),r+s)$, there exists a real polynomial $Q_n(x)$ of degree $\leq n$ such that

$$L(Q_n)\geq 0 \text{ throughout } I,\qquad (3.6)$$

and

$$\left\| f^{(k)} - Q_n^{(k)} \right\|_\infty \leq \frac{C}{n^{r-v}} \omega_s \left(f^{(r)}, \frac{1}{n} \right), \quad 0 \leq k \leq h. \tag{3.7}$$

Moreover, we get

$$\left\| f^{(k)} - Q_n^{(k)} \right\|_\infty \leq \frac{C}{n^{r-k}} \omega_s \left(f^{(r)}, \frac{1}{n} \right), \quad h+1 \leq k \leq r, \tag{3.8}$$

were C is a constant independent of f and n.

In this chapter we extend Theorem 3.3 to the fractional level. Indeed here L is replaced by L^*, a linear left Caputo fractional differential operator. Now the monotonicity property is only true on the critical interval $[0, 1]$. Simultaneous and fractional convergence remains true on all of I.

We are also inspired by [1].

We make

Definition 3.1. ([6], p. 50) Let $\alpha > 0$ and $\lceil \alpha \rceil = m$, ($\lceil \cdot \rceil$ ceiling of the number). Consider $f \in C^m([-1, 1])$. We define the left Caputo fractional derivative of f of order α as follows:

$$\left(D_{*-1}^{\alpha} f \right)(x) = \frac{1}{\Gamma(m-\alpha)} \int_{-1}^{x} (x-t)^{m-\alpha-1} f^{(m)}(t)\, dt, \tag{3.9}$$

for any $x \in [-1, 1]$, where Γ is the gamma function.

We set

$$D_{*-1}^{0} f(x) = f(x),$$

$$D_{*-1}^{m} f(x) = f^{(m)}(x), \ \forall\, x \in [-1, 1]. \tag{3.10}$$

3.2 Main Result

We present

Theorem 3.4. *Let h, v, r be integers, $1 \leq h \leq v \leq r$ and let $f \in C^r([-1, 1])$, with $f^{(r)}$ having modulus of smoothness $\omega_s\left(f^{(r)}, \delta\right)$ there, $s \geq 1$. Let $\alpha_j(x)$, $j = h, h+1, ..., v$ be real functions, defined and bounded on $[-1, 1]$ and suppose $\alpha_h(x)$ is either $\geq \alpha > 0$ or $\leq \beta < 0$ on $[0, 1]$. Let the real numbers $\alpha_0 = 0 < \alpha_1 \leq 1 < \alpha_2 \leq 2 < ... < \alpha_r \leq r$. Here $D_{*-1}^{\alpha_j} f$ stands for the left Caputo fractional derivative of f of order α_j anchored at -1. Consider the linear left fractional differential operator*

$$L^* := \sum_{j=h}^{k} \alpha_j(x) \left[D_{*-1}^{\alpha_j} \right] \tag{3.11}$$

and suppose, throughout $[0, 1]$,

$$L^*(f) \geq 0. \tag{3.12}$$

Then, for any $n \in \mathbb{N}$ such that $n \geq \max(4(r+1), r+s)$, there exists a real polynomial $Q_n(x)$ of degree $\leq n$ such that

$$L^*(Q_n) \geq 0 \quad \text{throughout} \quad [0,1], \tag{3.13}$$

and

$$\sup_{-1\leq x\leq 1} \left|\left(D_{*-1}^{\alpha_j} f\right)(x) - \left(D_{*-1}^{\alpha_j} Q_n\right)(x)\right|$$
$$\leq \frac{2^{j-\alpha_j}}{\Gamma(j-\alpha_j+1)} \frac{C_{r,s}}{n^{r-j}} \omega_s\left(f^{(r)}, \frac{1}{n}\right), \tag{3.14}$$

$j = h+1, ..., r$; $C_{r,s}$ is a constant independent of f and n.

Set

$$l_j :\equiv \sup_{x\in[-1,1]} \left|\alpha_h^{-1}(x)\,\alpha_j(x)\right|, \quad h \leq j \leq v. \tag{3.15}$$

When $j = 1, ..., h$ we derive

$$\sup_{-1\leq x\leq 1} \left|\left(D_{*-1}^{\alpha_j} f\right)(x) - \left(D_{*-1}^{\alpha_j} Q_n\right)(x)\right| \leq \frac{C_{r,s}}{n^{r-v}} \omega_s\left(f^{(r)}, \frac{1}{n}\right) \cdot$$

$$\left[\left(\sum_{\tau=h}^{v} l_\tau \frac{2^{\tau-\alpha_\tau}}{\Gamma(\tau-\alpha_\tau+1)}\right)\left(\sum_{\lambda=0}^{h-j} \frac{2^{h-\alpha_j-\lambda}}{\lambda!\,\Gamma(h-\alpha_j-\lambda+1)}\right) + \frac{2^{j-\alpha_j}}{\Gamma(j-\alpha_j+1)}\right]. \tag{3.16}$$

Finally it holds

$$\sup_{-1\leq x\leq 1} |f(x) - Q_n(x)|$$
$$\leq \frac{C_{r,s}}{n^{r-v}} \omega_s\left(f^{(r)}, \frac{1}{n}\right)\left[\frac{1}{h!}\sum_{\tau=h}^{v} l_\tau \frac{2^{\tau-\alpha_\tau}}{\Gamma(\tau-\alpha_\tau+1)} + 1\right]. \tag{3.17}$$

Proof. Here let Q_n as in Corollary 3.1.

Let $\alpha_j > 0$, $j = 1, ..., r$, such that $0 < \alpha_1 \leq 1 < \alpha_2 \leq 2 < \alpha_3 \leq 3... < ... < \alpha_r \leq r$. That is $\lceil \alpha_j \rceil = j$, $j = 1, ..., r$.

We consider the left Caputo fractional derivatives

$$\left(D_{*-1}^{\alpha_j} f\right)(x) = \frac{1}{\Gamma(j-\alpha_j)} \int_{-1}^{x} (x-t)^{j-\alpha_j-1} f^{(j)}(t)\, dt, \tag{3.18}$$

and

$$\left(D_{*-1}^{j} f\right)(x) = f^{(j)}(x),$$

and

$$\left(D_{*-1}^{\alpha_j} Q_n\right)(x) = \frac{1}{\Gamma(j-\alpha_j)} \int_{-1}^{x} (x-t)^{j-\alpha_j-1} Q_n^{(j)}(t)\, dt, \tag{3.19}$$

$$\left(D_{*-1}^{j} Q_n\right)(x) = Q_n^{(j)}(x); \; j = 1, ..., r.$$

We notice that

$$\begin{aligned}
&\left|\left(D_{*-1}^{\alpha_j} f\right)(x)-\left(D_{*-1}^{\alpha_j} Q_n\right)(x)\right| \\
&=\frac{1}{\Gamma\left(j-\alpha_j\right)}\left|\int_{-1}^{x}(x-t)^{j-\alpha_j-1} f^{(j)}(t)\, dt-\int_{-1}^{x}(x-t)^{j-\alpha_j-1} Q_n^{(j)}(t)\, dt\right| \qquad (3.20)\\
&=\frac{1}{\Gamma\left(j-\alpha_j\right)}\left|\int_{-1}^{x}(x-t)^{j-\alpha_j-1}\left(f^{(j)}(t)-Q_n^{(j)}(t)\right) dt\right| \\
&\leq \frac{1}{\Gamma\left(j-\alpha_j\right)} \int_{-1}^{x}(x-t)^{j-\alpha_j-1}\left|f^{(j)}(t)-Q_n^{(j)}(t)\right| dt \qquad (3.21)\\
&\overset{(3.3)}{\leq} \frac{1}{\Gamma\left(j-\alpha_j\right)}\left(\int_{-1}^{x}(x-t)^{j-\alpha_j-1} dt\right) \frac{C_{r,s}}{n^{r-j}} \omega_s\left(f^{(r)}, \frac{1}{n}\right) \\
&=\frac{1}{\Gamma\left(j-\alpha_j\right)} \frac{(x+1)^{j-\alpha_j}}{\left(j-\alpha_j\right)} \frac{C_{r,s}}{n^{r-j}} \omega_s\left(f^{(r)}, \frac{1}{n}\right) \qquad (3.22)\\
&=\frac{(x+1)^{j-\alpha_j}}{\Gamma\left(j-\alpha_j+1\right)} \frac{C_{r,s}}{n^{r-j}} \omega_s\left(f^{(r)}, \frac{1}{n}\right) \\
&\leq \frac{2^{j-\alpha_j}}{\Gamma\left(j-\alpha_j+1\right)} \frac{C_{r,s}}{n^{r-j}} \omega_s\left(f^{(r)}, \frac{1}{n}\right).
\end{aligned}$$

We proved for any $x \in [-1,1]$ that

$$\left|\left(D_{*-1}^{\alpha_j} f\right)(x)-\left(D_{*-1}^{\alpha_j} Q_n\right)(x)\right| \leq \frac{2^{j-\alpha_j}}{\Gamma\left(j-\alpha_j+1\right)} \frac{C_{r,s}}{n^{r-j}} \omega_s\left(f^{(r)}, \frac{1}{n}\right). \qquad (3.23)$$

Hence it holds

$$\sup_{-1\leq x\leq 1}\left|\left(D_{*-1}^{\alpha_j} f\right)(x)-\left(D_{*-1}^{\alpha_j} Q_n\right)(x)\right| \leq \frac{2^{j-\alpha_j}}{\Gamma\left(j-\alpha_j+1\right)} \frac{C_{r,s}}{n^{r-j}} \omega_s\left(f^{(r)}, \frac{1}{n}\right), \qquad (3.24)$$

$j = 0, 1, ..., r$.

Above we set $D_{*-1}^{0} f(x) = f(x)$, $D_{*-1}^{0} Q_n(x) = Q_n(x)$, $\forall\, x \in [-1,1]$, and $\alpha_0 = 0$, i.e. $\lceil \alpha_0 \rceil = 0$.

Set also

$$\rho_n := C_{r,s}\omega_s\left(f^{(r)}, \frac{1}{n}\right)\left(\sum_{j=h}^{v} l_j \frac{2^{j-\alpha_j}}{\Gamma\left(j-\alpha_j+1\right)} n^{j-r}\right). \qquad (3.25)$$

I. Suppose, throughout $[0,1]$, $\alpha_h(x) \geq \alpha > 0$. Let $Q_n(x)$, $x \in [-1,1]$, be a real polynomial of degree $\leq n$ so that

$$\max_{-1\leq x\leq 1}\left|D_{*-1}^{\alpha_j}\left(f(x)+\rho_n \frac{x^h}{h!}\right)-\left(D_{*-1}^{\alpha_j} Q_n\right)(x)\right|$$

$$\leq \frac{2^{j-\alpha_j}}{\Gamma(j-\alpha_j+1)} \frac{C_{r,s}}{n^{r-j}} \omega_s \left(f^{(r)}, \frac{1}{n}\right), \tag{3.26}$$

$j = 0, 1, ..., r$.

When $j = h+1, ..., r$, then

$$\max_{-1 \leq x \leq 1} \left| \left(D_{*-1}^{\alpha_j} f\right)(x) - \left(D_{*-1}^{\alpha_j} Q_n\right)(x) \right|$$

$$\leq \frac{2^{j-\alpha_j}}{\Gamma(j-\alpha_j+1)} \frac{C_{r,s}}{n^{r-j}} \omega_s \left(f^{(r)}, \frac{1}{n}\right), \tag{3.27}$$

proving (3.14).

For $j = 1, ..., h$ we get

$$D_{*-1}^{\alpha_j} \left(\frac{x^h}{h!}\right) = \frac{1}{\Gamma(j-\alpha_j)} \int_{-1}^{x} (x-t)^{j-\alpha_j-1} \frac{t^{h-j}}{(h-j)!} dt \tag{3.28}$$

(we see that $t = t+1-1$, and

$$t^{h-j} = ((t+1)-1)^{h-j} = \sum_{\lambda=0}^{h-j} \binom{h-j}{\lambda} (t+1)^{h-j-\lambda} (-1)^{\lambda})$$

$$= \frac{1}{(h-j)!\Gamma(j-\alpha_j)} \sum_{\lambda=0}^{h-j} (-1)^{\lambda} \binom{h-j}{\lambda} \int_{-1}^{x} (x-t)^{j-\alpha_j-1} (t+1)^{h-j-\lambda+1-1} dt$$

$$= \frac{1}{(h-j)!\Gamma(j-\alpha_j)} \sum_{\lambda=0}^{h-j} (-1)^{\lambda} \frac{(h-j)!}{\lambda!(h-j-\lambda)!} \cdot$$

$$\frac{\Gamma(j-\alpha_j)\Gamma(h-j-\lambda+1)}{\Gamma(h-\alpha_j-\lambda+1)} (x+1)^{h-\alpha_j-\lambda}$$

$$= \sum_{\lambda=0}^{h-j} \frac{(-1)^{\lambda}}{\lambda!\Gamma(h-\alpha_j-\lambda+1)} (x+1)^{h-\alpha_j-\lambda}. \tag{3.29}$$

Hence for $j = 1, ..., h$ we found that

$$D_{*-1}^{\alpha_j} \left(\frac{x^h}{h!}\right) = \sum_{\lambda=0}^{h-j} \frac{(-1)^{\lambda} (x+1)^{h-\alpha_j-\lambda}}{\lambda!\Gamma(h-\alpha_j-\lambda+1)}. \tag{3.30}$$

Therefore we get from (3.26) that

$$\max_{-1 \leq x \leq 1} \left| \left(D_{*-1}^{\alpha_j} f\right)(x) + \rho_n \left(\sum_{\lambda=0}^{h-j} \frac{(-1)^{\lambda} (x+1)^{h-\alpha_j-\lambda}}{\lambda!\Gamma(h-\alpha_j-\lambda+1)}\right) - \left(D_{*-1}^{\alpha_j} Q_n\right)(x) \right| \tag{3.31}$$

$$\leq \frac{2^{j-\alpha_j}}{\Gamma(j-\alpha_j+1)} \frac{C_{r,s}}{n^{r-j}} \omega_s \left(f^{(r)}, \frac{1}{n}\right),$$

$j = 1, ..., h$.

Therefore we get for $j = 1, ..., h$, that

$$\max_{-1\leq x\leq 1} \left| \left(D_{*-1}^{\alpha_j} f\right)(x) - \left(D_{*-1}^{\alpha_j} Q_n\right)(x) \right|$$

$$\leq \rho_n \left(\sum_{\lambda=0}^{h-j} \frac{2^{h-\alpha_j-\lambda}}{\lambda! \Gamma(h-\alpha_j-\lambda+1)} \right) + \frac{2^{j-\alpha_j}}{\Gamma(j-\alpha_j+1)} \frac{C_{r,s}}{n^{r-j}} \omega_s \left(f^{(r)}, \frac{1}{n} \right) \tag{3.32}$$

$$= C_{r,s} \omega_s \left(f^{(r)}, \frac{1}{n} \right) \left(\sum_{\overline{j}=h}^{k} l_{\overline{j}} \frac{2^{\overline{j}-\alpha_{\overline{j}}}}{\Gamma\left(\overline{j}-\alpha_{\overline{j}}+1\right)} n^{\overline{j}-r} \right)$$

$$\cdot \left(\sum_{\lambda=0}^{h-j} \frac{2^{h-\alpha_j-\lambda}}{\lambda! \Gamma(h-\alpha_j-\lambda+1)} \right) + \frac{2^{j-\alpha_j}}{\Gamma(j-\alpha_j+1)} \frac{C_{r,s}}{n^{r-j}} \omega_s \left(f^{(r)}, \frac{1}{n} \right)$$

$$= C_{r,s} \omega_s \left(f^{(r)}, \frac{1}{n} \right) \left[\left(\sum_{\overline{j}=h}^{k} l_{\overline{j}} \frac{2^{\overline{j}-\alpha_{\overline{j}}}}{\Gamma\left(\overline{j}-\alpha_{\overline{j}}+1\right)} \frac{1}{n^{r-\overline{j}}} \right) \right. \tag{3.33}$$

$$\left. \cdot \left(\sum_{\lambda=0}^{h-j} \frac{2^{h-\alpha_j-\lambda}}{\lambda! \Gamma(h-\alpha_j-\lambda+1)} \right) + \frac{2^{j-\alpha_j}}{\Gamma(j-\alpha_j+1)} \frac{1}{n^{r-j}} \right]$$

$$\leq C_{r,s} \omega_s \left(f^{(r)}, \frac{1}{n} \right) \frac{1}{n^{r-v}} \left[\left(\sum_{\overline{j}=h}^{v} l_{\overline{j}} \frac{2^{\overline{j}-\alpha_{\overline{j}}}}{\Gamma\left(\overline{j}-\alpha_{\overline{j}}+1\right)} \right) \right. \tag{3.34}$$

$$\left. \cdot \left(\sum_{\lambda=0}^{h-j} \frac{2^{h-\alpha_j-\lambda}}{\lambda! \Gamma(h-\alpha_j-\lambda+1)} \right) + \frac{2^{j-\alpha_j}}{\Gamma(j-\alpha_j+1)} \right].$$

Hence for $j = 1, ..., h$ we derived (3.16):

$$\max_{-1\leq x\leq 1} \left| \left(D_{*-1}^{\alpha_j} f\right)(x) - \left(D_{*-1}^{\alpha_j} Q_n\right)(x) \right| \leq \frac{C_{r,s}}{n^{r-v}} \omega_s \left(f^{(r)}, \frac{1}{n} \right)$$

$$\cdot \left[\left(\sum_{\tau=h}^{v} l_\tau \frac{2^{\tau-\alpha_\tau}}{\Gamma(\tau-\alpha_\tau+1)} \right) \left(\sum_{\lambda=0}^{h-j} \frac{2^{h-\alpha_j-\lambda}}{\lambda! \Gamma(h-\alpha_j-\lambda+1)} \right) + \frac{2^{j-\alpha_j}}{\Gamma(j-\alpha_j+1)} \right]. \tag{3.35}$$

From (3.26) when $j = 0$ we obtain

$$\max_{-1\leq x\leq 1} \left| f(x) + \rho_n \frac{x^h}{h!} - Q_n(x) \right| \leq \frac{C_{r,s}}{n^r} \omega_s \left(f^{(r)}, \frac{1}{n} \right). \tag{3.36}$$

And

$$\max_{-1\leq x\leq 1} \left| f(x) - Q_n(x) \right| \leq \frac{\rho_n}{h!} + \frac{C_{r,s}}{n^r} \omega_s \left(f^{(r)}, \frac{1}{n} \right) \tag{3.37}$$

$$= \frac{C_{r,s}}{h!}\omega_s\left(f^{(r)},\frac{1}{n}\right)\left(\sum_{\tau=h}^{v} l_\tau \frac{2^{\tau-\alpha_\tau}}{\Gamma(\tau-\alpha_\tau+1)} n^{\tau-r}\right)$$

$$+\frac{C_{r,s}}{n^r}\omega_s\left(f^{(r)},\frac{1}{n}\right)$$

$$= C_{r,s}\omega_s\left(f^{(r)},\frac{1}{n}\right)\left[\frac{1}{h!}\sum_{\tau=h}^{v} l_\tau \frac{2^{\tau-\alpha_\tau}}{\Gamma(\tau-\alpha_\tau+1)\, n^{r-\tau}} + \frac{1}{n^r}\right]$$

$$\leq \frac{C_{r,s}}{n^{r-v}}\omega_s\left(f^{(r)},\frac{1}{n}\right)\left[\frac{1}{h!}\sum_{\tau=h}^{k} l_\tau \frac{2^{\tau-\alpha_\tau}}{\Gamma(\tau-\alpha_\tau+1)} + 1\right], \tag{3.38}$$

that is proving (3.17).

Also if $0 \leq x \leq 1$, then

$$\alpha_h^{-1}(x)\, L^*(Q_n(x)) = \alpha_h^{-1}(x)\, L^*(f(x)) + \rho_n \frac{(x+1)^{h-\alpha_h}}{\Gamma(h-\alpha_h+1)} \tag{3.39}$$

$$+\sum_{j=h}^{v} \alpha_h^{-1}(x)\,\alpha_j(x)\left[D_{*-1}^{\alpha_j} Q_n(x) - D_{*-1}^{\alpha_j} f(x) - \frac{\rho_n}{h!} D_{*-1}^{\alpha_j} x^h\right]$$

$$\overset{(3.26)}{\geq} \rho_n \frac{(x+1)^{h-\alpha_h}}{\Gamma(h-\alpha_h+1)} - \left(\sum_{j=h}^{v} l_j \frac{2^{j-\alpha_j}}{\Gamma(j-\alpha_j+1)} \frac{C_{r,s}}{n^{r-j}}\omega_s\left(f^{(r)},\frac{1}{n}\right)\right)$$

$$= \rho_n \frac{(x+1)^{h-\alpha_h}}{\Gamma(h-\alpha_h+1)} - \rho_n = \rho_n\left[\frac{(x+1)^{h-\alpha_h}}{\Gamma(h-\alpha_h+1)} - 1\right] \tag{3.40}$$

$$= \rho_n\left[\frac{(x+1)^{h-\alpha_h} - \Gamma(h-\alpha_h+1)}{\Gamma(h-\alpha_h+1)}\right] \geq \rho_n\left[\frac{1-\Gamma(h-\alpha_h+1)}{\Gamma(h-\alpha_h+1)}\right] \geq 0. \tag{3.41}$$

Explanation: We know that $\Gamma(1) = 1$, $\Gamma(2) = 1$, and Γ is convex and positive on $(0,\infty)$. Here $0 \leq h-\alpha_h < 1$ and $1 \leq h-\alpha_h+1 < 2$. Thus $\Gamma(h-\alpha_h+1) \leq 1$ and $1-\Gamma(h-\alpha_h+1) \geq 0$. Hence $L^*(Q_n(x)) \geq 0$, $x \in [0,1]$.

II. Suppose on $[0,1]$ that $\alpha_h(x) \leq \beta < 0$. Let $Q_n(x)$, $x \in [-1,1]$, be a real polynomial of degree $\leq n$ so that

$$\max_{-1\leq x\leq 1}\left|D_{*-1}^{\alpha_j}\left(f(x) - \rho_n \frac{x^h}{h!}\right) - \left(D_{*-1}^{\alpha_j} Q_n\right)(x)\right|$$

$$\leq \frac{2^{j-\alpha_j}}{\Gamma(j-\alpha_j+1)} \frac{C_{r,s}}{n^{r-j}}\omega_s\left(f^{(r)},\frac{1}{n}\right), \tag{3.42}$$

$j = 0, 1, ..., r$.

Similarly we obtain again inequalities of convergence, see (3.14), (3.16) and (3.17).

Also if $0 \leq x \leq 1$, then

$$\alpha_h^{-1}(x) L^*(Q_n(x)) = \alpha_h^{-1}(x) L^*(f(x)) - \rho_n \frac{(x+1)^{h-\alpha_h}}{\Gamma(h-\alpha_h+1)} \tag{3.43}$$

$$+ \sum_{j=h}^{v} \alpha_h^{-1}(x) \alpha_j(x) \left[D_{*-1}^{\alpha_j} Q_n(x) - D_{*-1}^{\alpha_j} f(x) + \frac{\rho_n}{h!} \left(D_{*-1}^{\alpha_j} x^h \right) \right]$$

$$\overset{(3.42)}{\leq} -\rho_n \frac{(x+1)^{h-\alpha_h}}{\Gamma(h-\alpha_h+1)} + \sum_{j=h}^{v} l_j \frac{2^{j-\alpha_j}}{\Gamma(j-\alpha_j+1)} \frac{C_{r,s}}{n^{r-j}} \omega_s \left(f^{(r)}, \frac{1}{n} \right)$$

$$= \rho_n \left(1 - \frac{(x+1)^{h-\alpha_h}}{\Gamma(h-\alpha_h+1)} \right) = \rho_n \left(\frac{\Gamma(h-\alpha_h+1) - (x+1)^{h-\alpha_h}}{\Gamma(h-\alpha_h+1)} \right) \tag{3.44}$$

$$\leq \rho_n \left(\frac{1 - (x+1)^{h-\alpha_h}}{\Gamma(h-\alpha_h+1)} \right) \leq 0, \tag{3.45}$$

and hence on $[0,1]$ again holds $L^*(Q_n(x)) \geq 0$. □

Remark 3.1. (to Theorem 3.4) Suppose that $\alpha_j(x)$, $j = h, h+1, ..., v$ are continuous functions on $[-1,1]$, and we have on $[0,1]$ only $L^*(f) > 0$. Relax the condition $\alpha_h(x)$ is either $\geq \alpha > 0$ or $\leq \beta < 0$ on $[0,1]$. Let Q_n be the polynomial of degree $\leq n$ corresponding to f from (3.24).

Then $D_{*-1}^{\alpha_j} Q_n$ converges uniformly to $D_{*-1}^{\alpha_j} f$ at a higher rate given by inequality (3.24), in particular for $0 \leq j \leq h$. Moreover, because $L^*(Q_n)$ converges uniformly to $L^*(f)$ on $[-1,1]$, $L^*(Q_n) > 0$ on $[0,1]$ for sufficiently large n.

Bibliography

1. G.A. Anastassiou, *Bivariate monotone approximation*, Proc. Amer. Math. Soc. 112 (4) (1991), 959-963.
2. G.A. Anastassiou, *Higher order monotone approximation with linear differential operators*, Indian J. Pure Appl. Math. 24 (4) (1993), 263-266.
3. G.A. Anastassiou, O. Shisha, *Monotone approximation with linear differential operators*, J. Approx. Theory 44 (1985), 391-393.
4. G.A. Anastassiou, *Univariate left fractional polynomial high order monotone approximation,* Bulletin Korean Math. Soc., accepted 2014.
5. H.H. Gonska, E. Hinnemann, *Pointwise estimated for approximation by algebraic polynomials*, Acta Math. Hungar. 46 (1985), 243-254.
6. K. Diethelm, *The Analysis of Fractional Differential Equations*, Lecture Notes in Mathematics, Vol. 2004, 1st edition, Springer, New York, Heidelberg, 2010.
7. O. Shisha, *Monotone approximation*, Pacific J. Math. 15 (1965), 667-671.

Chapter 4

Univariate Right Fractional Polynomial High Order Monotone Approximation Theory

Let $f \in C^r([-1,1])$, $r \geq 0$ and let L^* be a linear right fractional differential operator such that $L^*(f) \geq 0$ throughout $[-1,0]$. We can find a sequence of polynomials Q_n of degree $\leq n$ such that $L^*(Q_n) \geq 0$ over $[-1,0]$, furthermore f is approximated right fractionally and simultaneously by Q_n on $[-1,1]$. The degree of these restricted approximations is given via inequalities using a higher order modulus of smoothness for $f^{(r)}$. It follows [4].

4.1 Introduction

The topic of monotone approximation started in [7] has become a major trend in approximation theory. A typical problem in this subject is: given a positive integer k, approximate a given function whose kth derivative is ≥ 0 by polynomials having this property.

In [3] the authors replaced the kth derivative with a linear differential operator of order k. We mention this motivating result.

Theorem 4.1. *Let h, k, p be integers, $0 \leq h \leq k \leq p$ and let f be a real function, $f^{(p)}$ continuous in $[-1,1]$ with modulus of continuity $\omega_1(f^{(p)}, x)$ there. Let $a_j(x)$, $j = h, h+1, ..., k$ be real functions, defined and bounded on $[-1,1]$ and assume $a_h(x)$ is either $\geq$ some number $\alpha > 0$ or $\leq$ some number $\beta < 0$ throughout $[-1,1]$. Consider the operator*

$$L = \sum_{j=h}^{k} a_j(x) \left[\frac{d^j}{dx^j} \right]$$

and suppose, throughout $[-1,1]$,

$$L(f) \geq 0. \tag{4.1}$$

Then, for every integer $n \geq 1$, there is a real polynomial $Q_n(x)$ of degree $\leq n$ such that

$$L(Q_n) \geq 0 \text{ throughout } [-1,1]$$

and

$$\max_{-1\le x\le 1} |f(x) - Q_n(x)| \le C n^{k-p} \omega_1 \left(f^{(p)}, \frac{1}{n}\right),$$

where C is independent of n and f.

We use also the notation $I = [-1,1]$.
We would like to mention

Theorem 4.2. *(Gonska and Hinnemann [6]). Let $r \ge 0$ and $s \ge 1$. Then there exists a sequence $Q_n = Q_n^{(r,s)}$ of linear polynomial operators mapping of $C^r(I)$ into P_n (space of polynomials of degree $\le n$), such that for all $f \in C^r(I)$, all $|x| \le 1$ and all $n \ge \max(4(r+1), r+s)$ we have*

$$\left|f^{(k)}(x) - (Q_n f)^{(k)}(x)\right| \le M_{r,s} (\Delta_n(x))^{r-k} \omega_s \left(f^{(r)}, \Delta_n(x)\right), \quad 0 \le k \le r, \tag{4.2}$$

where $\Delta_n(x) = \frac{\sqrt{1-x^2}}{n} + \frac{1}{n^2}$, and $M_{r,s}$ is a constant independent of f, x, and n. Above ω_s is the usual modulus of smoothness of order s with respect to the supremum norm.

Theorem 4.2 implies the useful

Corollary 4.1. *([2]) Let $r \ge 0$ and $s \ge 1$. Then there exists a sequence $Q_n = Q_n^{(r,s)}$ of linear polynomial operators mapping of $C^r(I)$ into P_n, such that for all $f \in C^r(I)$ and all $n \ge \max(4(r+1), r+s)$ we have*

$$\left\|f^{(k)} - (Q_n f)^{(k)}\right\|_\infty \le \frac{C_{r,s}}{n^{r-k}} \omega_s \left(f^{(r)}, \frac{1}{n}\right), \quad k = 0, 1, ..., r, \tag{4.3}$$

where $C_{r,s}$ is a constant independent of f and n.

In [2] we proved the following

Theorem 4.3. *Let h, v, r be integers, $0 \le h \le v \le r$ and let $f \in C^r(I)$, with $f^{(r)}$ having modulus of smoothness $\omega_s\left(f^{(r)}, \delta\right)$ there, $s \ge 1$. Let $\alpha_j(x)$, $j = h, h+1, ..., v$ be real functions, defined and bounded on I and suppose α_h is either $\ge \alpha > 0$ or $\le \beta < 0$ throughout I. Take the operator*

$$L = \sum_{j=h}^{v} \alpha_j(x) \left[\frac{d^j}{dx^j}\right] \tag{4.4}$$

and assume, throughout I,

$$L(f) \ge 0. \tag{4.5}$$

Then for every integer $n \ge \max(4(r+1), r+s)$, there exists a real polynomial $Q_n(x)$ of degree $\le n$ such that

$$L(Q_n) \ge 0 \text{ throughout } I, \tag{4.6}$$

and

$$\left\| f^{(k)} - Q_n^{(k)} \right\|_\infty \leq \frac{C}{n^{r-v}} \omega_s \left(f^{(r)}, \frac{1}{n} \right), \quad 0 \leq k \leq h. \tag{4.7}$$

Moreover, we get

$$\left\| f^{(k)} - Q_n^{(k)} \right\|_\infty \leq \frac{C}{n^{r-k}} \omega_s \left(f^{(r)}, \frac{1}{n} \right), \quad h+1 \leq k \leq r, \tag{4.8}$$

were C is a constant independent of f and n.

In this chapter we extend Theorem 4.3 to the right fractional level. Indeed here L is replaced by L^*, a linear right Caputo fractional differential operator. Now the monotonicity property is only true on the critical interval $[-1, 0]$. Simultaneous and right fractional convergence remains true on all of I.

We are also inspired by [1].

We make

Definition 4.1. ([5]) Let $\alpha > 0$ and $\lceil \alpha \rceil = m$, ($\lceil \cdot \rceil$ ceiling of the number). Consider $f \in C^m([-1, 1])$. The right Caputo fractional derivative of f of order α anchored at 1 is given by

$$\left(D_{1-}^{\alpha} f \right)(x) = \frac{(-1)^m}{\Gamma(m-\alpha)} \int_x^1 (t-x)^{m-\alpha-1} f^{(m)}(t)\, dt, \tag{4.9}$$

for any $x \in [-1, 1]$, where Γ is the gamma function.

In particular

$$D_{1-}^{0} f(x) = f(x),$$

$$D_{1-}^{m} f(x) = (-1)^m f^{(m)}(x), \quad \forall\, x \in [-1, 1]. \tag{4.10}$$

4.2 Main Result

We present

Theorem 4.4. *Let h, v, r be integers, h is even, $1 \leq h \leq v \leq r$ and let $f \in C^r([-1, 1])$, with $f^{(r)}$ having modulus of smoothness $\omega_s\left(f^{(r)}, \delta\right)$ there, $s \geq 1$. Let $\alpha_j(x)$, $j = h, h+1, ..., v$ be real functions, defined and bounded on $[-1, 1]$ and suppose $\alpha_h(x)$ is either $\geq \alpha > 0$ or $\leq \beta < 0$ on $[-1, 0]$. Let the real numbers $\alpha_0 = 0 < \alpha_1 < 1 < \alpha_2 < 2 < ... < \alpha_r < r$. Consider the linear right fractional differential operator*

$$L^* := \sum_{j=h}^{k} \alpha_j(x) \left[D_{1-}^{\alpha_j} \right] \tag{4.11}$$

and suppose, throughout $[-1, 0]$,

$$L^*(f) \geq 0. \tag{4.12}$$

Then, for any $n \in \mathbb{N}$ such that $n \geq \max(4(r+1), r+s)$, there exists a real polynomial Q_n of degree $\leq n$ such that

$$L^*(Q_n) \geq 0 \ \ \textit{throughout} \ \ [-1,0], \tag{4.13}$$

and

$$\sup_{-1\leq x\leq 1} \left| \left(D_{1-}^{\alpha_j} f\right)(x) - \left(D_{1-}^{\alpha_j} Q_n\right)(x) \right|$$
$$\leq \frac{2^{j-\alpha_j}}{\Gamma(j-\alpha_j+1)} \frac{C_{r,s}}{n^{r-j}} \omega_s \left(f^{(r)}, \frac{1}{n} \right), \tag{4.14}$$

$j = h+1, ..., r$; $C_{r,s}$ is a constant independent of f and n.

Set

$$l_j :\equiv \sup_{x\in[-1,1]} \left| \alpha_h^{-1}(x)\, \alpha_j(x) \right|, \quad h \leq j \leq v. \tag{4.15}$$

When $j = 1, ..., h$ we derive

$$\sup_{-1\leq x\leq 1} \left| \left(D_{1-}^{\alpha_j} f\right)(x) - \left(D_{1-}^{\alpha_j} Q_n\right)(x) \right| \leq \frac{C_{r,s}}{n^{r-v}} \omega_s \left(f^{(r)}, \frac{1}{n} \right) \cdot$$

$$\left[\left(\sum_{\tau=h}^{v} l_\tau \frac{2^{\tau-\alpha_\tau}}{\Gamma(\tau-\alpha_\tau+1)} \right) \left(\sum_{\lambda=0}^{h-j} \frac{2^{h-\alpha_j-\lambda}}{\lambda!\,\Gamma(h-\alpha_j-\lambda+1)} \right) + \frac{2^{j-\alpha_j}}{\Gamma(j-\alpha_j+1)} \right]. \tag{4.16}$$

Finally it holds

$$\sup_{-1\leq x\leq 1} \left| f(x) - Q_n(x) \right|$$
$$\leq \frac{C_{r,s}}{n^{r-v}} \omega_s \left(f^{(r)}, \frac{1}{n} \right) \left[\frac{1}{h!} \sum_{\tau=h}^{v} l_\tau \frac{2^{\tau-\alpha_\tau}}{\Gamma(\tau-\alpha_\tau+1)} + 1 \right]. \tag{4.17}$$

Proof. We will prove that the polynomial Q_n described in Corollary 4.1 satisfies assertions of Theorem 4.4.

Let $\alpha_j > 0$, $j = 1, ..., r$, such that $0 < \alpha_1 < 1 < \alpha_2 < 2 < \alpha_3 < 3... < ... < \alpha_r < r$. That is $\lceil \alpha_j \rceil = j$, $j = 1, ..., r$.

We consider the right Caputo fractional derivatives

$$\left(D_{1-}^{\alpha_j} f\right)(x) = \frac{(-1)^j}{\Gamma(j-\alpha_j)} \int_x^1 (t-x)^{j-\alpha_j-1} f^{(j)}(t)\, dt, \tag{4.18}$$

$$\left(D_{1-}^{j} f\right)(x) = (-1)^j f^{(j)}(x),$$

and

$$\left(D_{1-}^{\alpha_j} Q_n\right)(x) = \frac{(-1)^j}{\Gamma(j-\alpha_j)} \int_x^1 (t\text{-}x)^{j-\alpha_j-1} Q_n^{(j)}(t)\, dt, \tag{4.19}$$

$$\left(D_{1-}^{j}Q_{n}\right)(x)=(-1)^{j}\,Q_{n}^{(j)}(x)\,;\ j=1,...,r.$$

We notice that

$$\left|\left(D_{1-}^{\alpha_j}f\right)(x)-\left(D_{1-}^{\alpha_j}Q_n\right)(x)\right|$$

$$=\frac{1}{\Gamma(j-\alpha_j)}\left|\int_x^1(t-x)^{j-\alpha_j-1}f^{(j)}(t)\,dt-\int_x^1(t-x)^{j-\alpha_j-1}Q_n^{(j)}(t)\,dt\right| \tag{4.20}$$

$$=\frac{1}{\Gamma(j-\alpha_j)}\left|\int_x^1(t-x)^{j-\alpha_j-1}\left(f^{(j)}(t)-Q_n^{(j)}(t)\right)dt\right|$$

$$\leq\frac{1}{\Gamma(j-\alpha_j)}\int_x^1(t-x)^{j-\alpha_j-1}\left|f^{(j)}(t)-Q_n^{(j)}(t)\right|dt \tag{4.21}$$

$$\overset{(4.3)}{\leq}\frac{1}{\Gamma(j-\alpha_j)}\left(\int_x^1(t-x)^{j-\alpha_j-1}dt\right)\frac{C_{r,s}}{n^{r-j}}\omega_s\left(f^{(r)},\frac{1}{n}\right)$$

$$=\frac{1}{\Gamma(j-\alpha_j)}\frac{(1-x)^{j-\alpha_j}}{(j-\alpha_j)}\frac{C_{r,s}}{n^{r-j}}\omega_s\left(f^{(r)},\frac{1}{n}\right) \tag{4.22}$$

$$=\frac{(1-x)^{j-\alpha_j}}{\Gamma(j-\alpha_j+1)}\frac{C_{r,s}}{n^{r-j}}\omega_s\left(f^{(r)},\frac{1}{n}\right)$$

$$\leq\frac{2^{j-\alpha_j}}{\Gamma(j-\alpha_j+1)}\frac{C_{r,s}}{n^{r-j}}\omega_s\left(f^{(r)},\frac{1}{n}\right).$$

We proved for any $x\in[-1,1]$ that

$$\left|\left(D_{1-}^{\alpha_j}f\right)(x)-\left(D_{1-}^{\alpha_j}Q_n\right)(x)\right|\leq\frac{2^{j-\alpha_j}}{\Gamma(j-\alpha_j+1)}\frac{C_{r,s}}{n^{r-j}}\omega_s\left(f^{(r)},\frac{1}{n}\right). \tag{4.23}$$

Hence for $j=0,1,...,r$ *we have*

$$\sup_{-1\leq x\leq 1}\left|\left(D_{1-}^{\alpha_j}f\right)(x)-\left(D_{1-}^{\alpha_j}Q_n\right)(x)\right|\leq\frac{2^{j-\alpha_j}}{\Gamma(j-\alpha_j+1)}\frac{C_{r,s}}{n^{r-j}}\omega_s\left(f^{(r)},\frac{1}{n}\right). \tag{4.24}$$

Above we set $D_{1-}^{0}f(x)=f(x)$, $D_{1-}^{0}Q_n(x)=Q_n(x)$, $\forall\,x\in[-1,1]$, and $\alpha_0=0$, i.e. $\lceil\alpha_0\rceil=0$.

Put

$$\rho_n:=C_{r,s}\omega_s\left(f^{(r)},\frac{1}{n}\right)\left(\sum_{j=h}^{v}l_j\frac{2^{j-\alpha_j}}{\Gamma(j-\alpha_j+1)}n^{j-r}\right). \tag{4.25}$$

I. Suppose, throughout $[-1,0]$, $\alpha_h(x)\geq\alpha>0$. Let $Q_n(x)$, $x\in[-1,1]$, be a real polynomial of degree $\leq n$ so that

$$\max_{-1\leq x\leq 1}\left|D_{1-}^{\alpha_j}\left(f(x)+\rho_n\frac{x^h}{h!}\right)-\left(D_{1-}^{\alpha_j}Q_n\right)(x)\right|$$

$$\leq \frac{2^{j-\alpha_j}}{\Gamma(j-\alpha_j+1)} \frac{C_{r,s}}{n^{r-j}} \omega_s \left(f^{(r)}, \frac{1}{n} \right), \tag{4.26}$$

$j = 0, 1, ..., r$.

When $j = h+1, ..., r$, then

$$\max_{-1 \leq x \leq 1} \left| \left(D_{1-}^{\alpha_j} f \right)(x) - \left(D_{1-}^{\alpha_j} Q_n \right)(x) \right|$$

$$\leq \frac{2^{j-\alpha_j}}{\Gamma(j-\alpha_j+1)} \frac{C_{r,s}}{n^{r-j}} \omega_s \left(f^{(r)}, \frac{1}{n} \right), \tag{4.27}$$

proving (4.14).

When $j = 1, ..., h$ we get

$$D_{1-}^{\alpha_j} \left(\frac{x^h}{h!} \right) = \frac{(-1)^j}{\Gamma(j-\alpha_j)} \int_x^1 (t-x)^{j-\alpha_j-1} \frac{t^{h-j}}{(h-j)!} dt \tag{4.28}$$

(we see that $t = t+1-1$, and $-t = 1-t-1$)

$$= \frac{(-1)^{j+h-j}}{(h-j)!\Gamma(j-\alpha_j)} \int_x^1 (-t)^{h-j} (t-x)^{j-\alpha_j-1} dt$$

$$= \frac{(-1)^h}{(h-j)!\Gamma(j-\alpha_j)} \int_x^1 (1-t-1)^{h-j} (t-x)^{j-\alpha_j-1} dt$$

(we see that $((1-t)-1)^{h-j} = \sum_{\lambda=0}^{h-j} \binom{h-j}{\lambda} (1-t)^{h-j-\lambda} (-1)^\lambda$)

$$= \frac{(-1)^h}{(h-j)!\Gamma(j-\alpha_j)} \sum_{\lambda=0}^{h-j} \binom{h-j}{\lambda} (-1)^\lambda \int_x^1 (1-t)^{(h-j-\lambda+1)-1} (t-x)^{(j-\alpha_j)-1} dt$$

$$= \frac{(-1)^h}{(h-j)!\Gamma(j-\alpha_j)} \sum_{\lambda=0}^{h-j} \frac{(h-j)!(-1)^\lambda}{\lambda!(h-j-\lambda)!}$$

$$\cdot \frac{\Gamma(h-j-\lambda+1)\Gamma(j-\alpha_j)}{\Gamma(h-\alpha_j-\lambda+1)} (1-x)^{h-\alpha_j-\lambda}$$

$$= (-1)^h \sum_{\lambda=0}^{h-j} \frac{(-1)^\lambda (1-x)^{h-\alpha_j-\lambda}}{\lambda!\Gamma(h-\alpha_j-\lambda+1)}. \tag{4.29}$$

Hence for $j = 1, ..., h$ we have that

$$D_{1-}^{\alpha_j} \left(\frac{x^h}{h!} \right) = (-1)^h \sum_{\lambda=0}^{h-j} \frac{(-1)^\lambda (1-x)^{h-\alpha_j-\lambda}}{\lambda!\Gamma(h-\alpha_j-\lambda+1)}. \tag{4.30}$$

Therefore we get from (4.26) that

$$\max_{-1 \leq x \leq 1} \left| \left(D_{1-}^{\alpha_j} f \right)(x) + \rho_n \left((-1)^h \sum_{\lambda=0}^{h-j} \frac{(-1)^\lambda (1-x)^{h-\alpha_j-\lambda}}{\lambda!\Gamma(h-\alpha_j-\lambda+1)} \right) - \left(D_{1-}^{\alpha_j} Q_n \right)(x) \right| \tag{4.31}$$

$$\leq \frac{2^{j-\alpha_j}}{\Gamma(j-\alpha_j+1)} \frac{C_{r,s}}{n^{r-j}} \omega_s \left(f^{(r)}, \frac{1}{n} \right),$$

$j = 1, ..., h$.

Therefore we get for $j = 1, ..., h$, that

$$\max_{-1 \leq x \leq 1} \left| \left(D_{1-}^{\alpha_j} f \right)(x) - \left(D_{1-}^{\alpha_j} Q_n \right)(x) \right|$$

$$\leq \rho_n \left(\sum_{\lambda=0}^{h-j} \frac{2^{h-\alpha_j-\lambda}}{\lambda! \Gamma(h-\alpha_j-\lambda+1)} \right) + \frac{2^{j-\alpha_j}}{\Gamma(j-\alpha_j+1)} \frac{C_{r,s}}{n^{r-j}} \omega_s \left(f^{(r)}, \frac{1}{n} \right) \tag{4.32}$$

$$= C_{r,s} \omega_s \left(f^{(r)}, \frac{1}{n} \right) \left(\sum_{\bar{j}=h}^{v} l_{\bar{j}} \frac{2^{\bar{j}-\alpha_{\bar{j}}}}{\Gamma\left(\bar{j}-\alpha_{\bar{j}}+1\right)} n^{\bar{j}-r} \right)$$

$$\cdot \left(\sum_{\lambda=0}^{h-j} \frac{2^{h-\alpha_j-\lambda}}{\lambda! \Gamma(h-\alpha_j-\lambda+1)} \right) + \frac{2^{j-\alpha_j}}{\Gamma(j-\alpha_j+1)} \frac{C_{r,s}}{n^{r-j}} \omega_s \left(f^{(r)}, \frac{1}{n} \right)$$

$$= C_{r,s} \omega_s \left(f^{(r)}, \frac{1}{n} \right) \left[\left(\sum_{\bar{j}=h}^{v} l_{\bar{j}} \frac{2^{\bar{j}-\alpha_{\bar{j}}}}{\Gamma\left(\bar{j}-\alpha_{\bar{j}}+1\right)} \frac{1}{n^{r-\bar{j}}} \right) \right. \tag{4.33}$$

$$\left. \cdot \left(\sum_{\lambda=0}^{h-j} \frac{2^{h-\alpha_j-\lambda}}{\lambda! \Gamma(h-\alpha_j-\lambda+1)} \right) + \frac{2^{j-\alpha_j}}{\Gamma(j-\alpha_j+1)} \frac{1}{n^{r-j}} \right]$$

$$\leq C_{r,s} \omega_s \left(f^{(r)}, \frac{1}{n} \right) \frac{1}{n^{r-v}} \left[\left(\sum_{\bar{j}=h}^{v} l_{\bar{j}} \frac{2^{\bar{j}-\alpha_{\bar{j}}}}{\Gamma\left(\bar{j}-\alpha_{\bar{j}}+1\right)} \right) \right. \tag{4.34}$$

$$\left. \cdot \left(\sum_{\lambda=0}^{h-j} \frac{2^{h-\alpha_j-\lambda}}{\lambda! \Gamma(h-\alpha_j-\lambda+1)} \right) + \frac{2^{j-\alpha_j}}{\Gamma(j-\alpha_j+1)} \right].$$

Hence for $j = 1, ..., h$ we derived (4.16):

$$\max_{-1 \leq x \leq 1} \left| \left(D_{1-}^{\alpha_j} f \right)(x) - \left(D_{1-}^{\alpha_j} Q_n \right)(x) \right| \leq \frac{C_{r,s}}{n^{r-v}} \omega_s \left(f^{(r)}, \frac{1}{n} \right)$$

$$\cdot \left[\left(\sum_{\tau=h}^{v} l_\tau \frac{2^{\tau-\alpha_\tau}}{\Gamma(\tau-\alpha_\tau+1)} \right) \left(\sum_{\lambda=0}^{h-j} \frac{2^{h-\alpha_j-\lambda}}{\lambda! \Gamma(h-\alpha_j-\lambda+1)} \right) + \frac{2^{j-\alpha_j}}{\Gamma(j-\alpha_j+1)} \right]. \tag{4.35}$$

From (4.26) when $j = 0$ we obtain

$$\max_{-1 \leq x \leq 1} \left| f(x) + \rho_n \frac{x^h}{h!} - Q_n(x) \right| \leq \frac{C_{r,s}}{n^r} \omega_s \left(f^{(r)}, \frac{1}{n} \right). \tag{4.36}$$

And

$$\max_{-1\leq x\leq 1}|f(x)-Q_n(x)|\leq \frac{\rho_n}{h!}+\frac{C_{r,s}}{n^r}\omega_s\left(f^{(r)},\frac{1}{n}\right) \tag{4.37}$$

$$=\frac{C_{r,s}}{h!}\omega_s\left(f^{(r)},\frac{1}{n}\right)\left(\sum_{\tau=h}^{v}l_\tau\frac{2^{\tau-\alpha_\tau}}{\Gamma(\tau-\alpha_\tau+1)}n^{\tau-r}\right)$$

$$+\frac{C_{r,s}}{n^r}\omega_s\left(f^{(r)},\frac{1}{n}\right)$$

$$=C_{r,s}\omega_s\left(f^{(r)},\frac{1}{n}\right)\left[\frac{1}{h!}\sum_{\tau=h}^{v}l_\tau\frac{2^{\tau-\alpha_\tau}}{\Gamma(\tau-\alpha_\tau+1)n^{r-\tau}}+\frac{1}{n^r}\right]$$

$$\leq\frac{C_{r,s}}{n^{r-v}}\omega_s\left(f^{(r)},\frac{1}{n}\right)\left[\frac{1}{h!}\sum_{\tau=h}^{v}l_\tau\frac{2^{\tau-\alpha_\tau}}{\Gamma(\tau-\alpha_\tau+1)}+1\right], \tag{4.38}$$

that is proving (4.17).

Also if $-1\leq x\leq 0$, then

$$\alpha_h^{-1}(x)L^*(Q_n(x))=\alpha_h^{-1}(x)L^*(f(x))+\rho_n\frac{(1-x)^{h-\alpha_h}}{\Gamma(h-\alpha_h+1)} \tag{4.39}$$

$$+\sum_{j=h}^{v}\alpha_h^{-1}(x)\alpha_j(x)\left[D_{1-}^{\alpha_j}Q_n(x)-D_{1-}^{\alpha_j}f(x)-\frac{\rho_n}{h!}D_{1-}^{\alpha_j}x^h\right]$$

$$\overset{(4.26)}{\geq}\rho_n\frac{(1-x)^{h-\alpha_h}}{\Gamma(h-\alpha_h+1)}-\left(\sum_{j=h}^{v}l_j\frac{2^{j-\alpha_j}}{\Gamma(j-\alpha_j+1)}\frac{C_{r,s}}{n^{r-j}}\omega_s\left(f^{(r)},\frac{1}{n}\right)\right)$$

$$=\rho_n\frac{(1-x)^{h-\alpha_h}}{\Gamma(h-\alpha_h+1)}-\rho_n=\rho_n\left[\frac{(1-x)^{h-\alpha_h}}{\Gamma(h-\alpha_h+1)}-1\right] \tag{4.40}$$

$$=\rho_n\left[\frac{(1-x)^{h-\alpha_h}-\Gamma(h-\alpha_h+1)}{\Gamma(h-\alpha_h+1)}\right]\geq\rho_n\left[\frac{1-\Gamma(h-\alpha_h+1)}{\Gamma(h-\alpha_h+1)}\right]\geq 0. \tag{4.41}$$

Explanation: We know that $\Gamma(1)=1$, $\Gamma(2)=1$, and Γ is convex and positive on $(0,\infty)$. Here $0\leq h-\alpha_h<1$ and $1\leq h-\alpha_h+1<2$. Thus $\Gamma(h-\alpha_h+1)\leq 1$ and $1-\Gamma(h-\alpha_h+1)\geq 0$. Hence $L^*(Q_n(x))\geq 0$, $x\in[-1,0]$.

II. Suppose on $[-1,0]$ that $\alpha_h(x)\leq\beta<0$. Let $Q_n(x)$, $x\in[-1,1]$, be a real polynomial of degree $\leq n$ so that

$$\max_{-1\leq x\leq 1}\left|D_{1-}^{\alpha_j}\left(f(x)-\rho_n\frac{x^h}{h!}\right)-\left(D_{1-}^{\alpha_j}Q_n\right)(x)\right|$$

$$\leq \frac{2^{j-\alpha_j}}{\Gamma(j-\alpha_j+1)} \frac{C_{r,s}}{n^{r-j}} \omega_s \left(f^{(r)}, \frac{1}{n}\right), \tag{4.42}$$

$j = 0, 1, ..., r$.

Similarly we obtain again inequalities of convergence, see (4.14), (4.16) and (4.17).

Also if $-1 \leq x \leq 0$, then

$$\alpha_h^{-1}(x) L^*(Q_n(x)) = \alpha_h^{-1}(x) L^*(f(x)) - \rho_n \frac{(1-x)^{h-\alpha_h}}{\Gamma(h-\alpha_h+1)} \tag{4.43}$$

$$+ \sum_{j=h}^{v} \alpha_h^{-1}(x) \alpha_j(x) \left[D_{1-}^{\alpha_j} Q_n(x) - D_{1-}^{\alpha_j} f(x) + \frac{\rho_n}{h!} \left(D_{1-}^{\alpha_j} x^h \right) \right]$$

$$\overset{(4.42)}{\leq} -\rho_n \frac{(1-x)^{h-\alpha_h}}{\Gamma(h-\alpha_h+1)} + \sum_{j=h}^{v} l_j \frac{2^{j-\alpha_j}}{\Gamma(j-\alpha_j+1)} \frac{C_{r,s}}{n^{r-j}} \omega_s \left(f^{(r)}, \frac{1}{n}\right)$$

$$= \rho_n \left(1 - \frac{(1-x)^{h-\alpha_h}}{\Gamma(h-\alpha_h+1)}\right) = \rho_n \left(\frac{\Gamma(h-\alpha_h+1) - (1-x)^{h-\alpha_h}}{\Gamma(h-\alpha_h+1)}\right) \tag{4.44}$$

$$\leq \rho_n \left(\frac{1-(1-x)^{h-\alpha_h}}{\Gamma(h-\alpha_h+1)}\right) \leq 0, \tag{4.45}$$

and hence on $[-1, 0]$ again holds $L^*(Q_n(x)) \geq 0$. □

Remark 4.1. (to Theorem 4.4) Suppose that $\alpha_j(x)$, $j = h, h+1, ..., v$ are continuous functions on $[-1, 1]$, and we have on $[-1, 0]$ only $L^*(f) > 0$. Relax the condition $\alpha_h(x)$ is either $\geq \alpha > 0$ or $\leq \beta < 0$ on $[-1, 0]$. Let Q_n be the polynomial of degree $\leq n$ corresponding to f from (4.24).

Then $D_{1-}^{\alpha_j} Q_n$ converges uniformly to $D_{1-}^{\alpha_j} f$ at a higher rate given by inequality (4.24), in particular for $0 \leq j \leq h$. Moreover, because $L^*(Q_n)$ converges uniformly to $L^*(f)$ on $[-1, 1]$, $L^*(Q_n) > 0$ on $[-1, 0]$ for sufficiently large n.

Bibliography

1. G.A. Anastassiou, *Bivariate monotone approximation*, Proc. Amer. Math. Soc. 112 (4) (1991), 959-963.
2. G.A. Anastassiou, *Higher order monotone approximation with linear differential operators*, Indian J. Pure Appl. Math. 24 (4) (1993), 263-266.
3. G.A. Anastassiou, O. Shisha, *Monotone approximation with linear differential operators*, J. Approx. Theory 44 (1985), 391-393.
4. G.A. Anastassiou, *Univariate right fractional polynomial high order monotone approximation*, Demonstratio Math., accepted 2014.
5. A.M.A. El-Sayed, M. Gaber, *On the finite Caputo and finite Riesz derivatives*, Electron. J. Theoret. Phys. 3 (12) (2006), 81-95.
6. H.H. Gonska, E. Hinnemann, *Pointwise estimated for approximation by algebraic polynomials*, Acta Math. Hungar. 46 (1985), 243-254.
7. O. Shisha, *Monotone approximation*, Pacific J. Math. 15 (1965), 667-671.

Chapter 5

Spline Left Fractional Monotone Approximation Theory Using Left Fractional Differential Operators

Let $f \in C^s([-1,1])$, $s \in \mathbb{N}$ and L^* be a linear left fractional differential operator such that $L^*(f) \geq 0$ on $[0,1]$. Then there exists a sequence Q_n, $n \in \mathbb{N}$ of polynomial splines with equally spaced knots of given fixed order such that $L^*(Q_n) \geq 0$ on $[0,1]$. Furthermore f is approximated with rates fractionally and simultaneously by Q_n in the uniform norm. This constrained fractional approximation on $[-1,1]$ is given via inequalities involving a higher modulus of smoothness of $f^{(s)}$. It follows [2].

5.1 Introduction

Let $[a,b] \subset \mathbb{R}$ and for $n \geq 1$ consider the partition Δ_n with points $x_{in} = a + i\left(\frac{b-a}{n}\right)$, $i = 0, 1, ..., n$. Hence $\overline{\Delta}_n \equiv \max_{1 \leq i \leq n}(x_{in} - x_{i-1,n}) = \frac{b-a}{n}$.

Let $S_m(\Delta_n)$ be the space of polynomial splines of order $m > 0$ with simple knots at the points x_{in}, $i = 1, ..., n-1$. Then there exists a linear operator Q_n : $Q_n \equiv Q_n(f)$, mapping $B[a,b]$: the space of bounded real valued functions f on $[a,b]$, into $S_m(\Delta_n)$ (see [5], p. 224, Theorem 6.18).

From the same reference [5], p. 227, Corollary 6.21, we get

Corollary 5.1. *Let $1 \leq \sigma \leq m$, $n \geq 1$. Then for all $f \in C^{\sigma-1}[a,b]$; $r = 0, ..., \sigma-1$,*

$$\left\| f^{(r)} - Q_n^{(r)} \right\|_\infty \leq C_1 \left(\frac{b-a}{n}\right)^{\sigma-r-1} \omega_{m-\sigma+1}\left(f^{(\sigma-1)}, \frac{b-a}{n}\right), \tag{5.1}$$

where C_1 depends only on m, $C_1 = C_1(m)$.

By denoting $C_2 = C_1 \max_{0 \leq r \leq \sigma-1}(b-a)^{\sigma-r-1}$ we obtain

Lemma 5.1. *([1]) Let $1 \leq \sigma \leq m$, $n \geq 1$. Then for all $f \in C^{\sigma-1}[a,b]$; $r = 0, ..., \sigma-1$,*

$$\left\| f^{(r)} - Q_n^{(r)} \right\|_\infty \leq \frac{C_2}{n^{\sigma-r-1}} \omega_{m-\sigma+1}\left(f^{(\sigma-1)}, \frac{b-a}{n}\right), \tag{5.2}$$

where C_2 depends only on m, σ and $b-a$. Here $\omega_{m-\sigma+1}$ is the usual modulus of smoothness of order $m-\sigma+1$.

We are motivated by

Theorem 5.1. *([1]) Let h, k, σ, m be integers, $0 \leq h \leq k \leq \sigma - 1$, $\sigma \leq m$ and let $f \in C^{\sigma-1}[a,b]$. Let $\alpha_j(x) \in B[a,b]$, $j = h, h+1, ..., k$ and suppose that $\alpha_h(x) \geq \alpha > 0$ or $\alpha_h(x) \leq \beta < 0$ for all $x \in [a,b]$. Take the linear differential operator*

$$L = \sum_{j=h}^{k} \alpha_j(x) \left[\frac{d^j}{dx^j}\right] \tag{5.3}$$

and assume, throughout $[a,b]$,

$$L(f) \geq 0. \tag{5.4}$$

Then, for every integer $n \geq 1$, there is a polynomial spline function $Q_n(x)$ of order m with simple knots at $\left\{a + i\left(\frac{b-a}{n}\right),\ i = 1, ..., n-1\right\}$ such that $L(Q_n) \geq 0$ throughout $[a,b]$ and

$$\left\| f^{(r)} - Q_n^{(r)} \right\|_\infty \leq \frac{C}{n^{\sigma-k-1}} \omega_{m-\sigma+1}\left(f^{(\sigma-1)}, \frac{b-a}{n}\right), \quad 0 \leq r \leq h. \tag{5.5}$$

Moreover, we find

$$\left\| f^{(r)} - Q_n^{(r)} \right\|_\infty \leq \frac{C}{n^{\sigma-r-1}} \omega_{m-\sigma+1}\left(f^{(\sigma-1)}, \frac{b-a}{n}\right), \quad h+1 \leq r \leq \sigma - 1, \tag{5.6}$$

where C is a constant independent of f and n. It depends only on m, σ, L, a, b.

Next we specialize on the case of $a = -1$, $b = 1$. That is working on $[-1,1]$. By Lemma 5.1 we get

Lemma 5.2. *Let $1 \leq \sigma \leq m$, $n \geq 1$. Then for all $f \in C^{\sigma-1}([-1,1])$; $j = 0, 1, ..., \sigma - 1$,*

$$\left\| f^{(j)} - Q_n^{(j)} \right\|_\infty \leq \frac{C_2}{n^{\sigma-j-1}} \omega_{m-\sigma+1}\left(f^{(\sigma-1)}, \frac{2}{n}\right), \tag{5.7}$$

where $C_2 := C_2(m,\sigma) := C_1(m)\, 2^{\sigma-1}$.

Since

$$\omega_{m-\sigma+1}\left(f^{(\sigma-1)}, \frac{2}{n}\right) \leq 2^{m-\sigma+1} \omega_{m-\sigma+1}\left(f^{(\sigma-1)}, \frac{1}{n}\right) \tag{5.8}$$

(see [3], p. 45), we get

Lemma 5.3. *Let $1 \leq \sigma \leq m$, $n \geq 1$. Then for all $f \in C^{\sigma-1}([-1,1])$; $j = 0, 1, ..., \sigma - 1$,*

$$\left\| f^{(j)} - Q_n^{(j)} \right\|_\infty \leq \frac{C_2^*}{n^{\sigma-j-1}} \omega_{m-\sigma+1}\left(f^{(\sigma-1)}, \frac{1}{n}\right), \tag{5.9}$$

where $C_2^* := C_2^*(m,\sigma) := C_1(m)\, 2^m$.

We use a lot in this chapter Lemma 5.3.

In this chapter we extend Theorem 5.1 over $[-1,1]$ to the fractional level. Indeed here L is replaced by L^*, a linear left Caputo fractional differential operator. Now the monotonicity property is only true on the critical interval $[0,1]$. Simultaneous fractional convergence remains true on all of $[-1,1]$.

We make

Definition 5.1. ([4], p. 50) Let $\alpha > 0$ and $\lceil \alpha \rceil = m$, ($\lceil \cdot \rceil$ ceiling of the number). Consider $f \in C^m([-1,1])$. We define the left Caputo fractional derivative of f of order α as follows:

$$\left(D_{*-1}^{\alpha} f\right)(x) = \frac{1}{\Gamma(m-\alpha)} \int_{-1}^{x} (x-t)^{m-\alpha-1} f^{(m)}(t)\, dt, \tag{5.10}$$

for any $x \in [-1,1]$, where Γ is the gamma function.

We set

$$D_{*-1}^{0} f(x) = f(x),$$

$$D_{*-1}^{m} f(x) = f^{(m)}(x), \ \ \forall\, x \in [-1,1]. \tag{5.11}$$

5.2 Main Result

We present

Theorem 5.2. *Let h, k, σ, m be integers, $1 \leq \sigma \leq m$, $n \in \mathbb{N}$, with $0 \leq h \leq k \leq \sigma - 2$ and let $f \in C^{\sigma-1}([-1,1])$, with $f^{(\sigma-1)}$ having modulus of smoothness $\omega_{m-\sigma+1}\left(f^{(\sigma-1)}, \delta\right)$ there, $\delta > 0$. Let $\alpha_j(x)$, $j = h, h+1, ..., k$ be real functions, defined and bounded on $[-1,1]$ and suppose $\alpha_h(x)$ is either $\geq \alpha > 0$ or $\leq \beta < 0$ on $[0,1]$. Let the real numbers $\alpha_0 = 0 < \alpha_1 \leq 1 < \alpha_2 \leq 2 < ... < \alpha_{\sigma-2} \leq \sigma - 2$. Here $D_{*-1}^{\alpha_j} f$ stands for the left Caputo fractional derivative of f of order α_j anchored at -1. Consider the linear left fractional differential operator*

$$L^* := \sum_{j=h}^{k} \alpha_j(x) \left[D_{*-1}^{\alpha_j}\right] \tag{5.12}$$

and suppose, throughout $[0,1]$, $L^*(f) \geq 0$.

Then, for every integer $n \geq 1$, there exists a polynomial spline function $Q_n(x)$ of order $m > 0$ with simple knots at $\left\{-1 + i\frac{2}{n},\ i = 1, ..., n-1\right\}$ such that $L^*(Q_n) \geq 0$ throughout $[0,1]$, and

$$\sup_{-1 \leq x \leq 1} \left| \left(D_{*-1}^{\alpha_j} f\right)(x) - \left(D_{*-1}^{\alpha_j} Q_n\right)(x) \right|$$

$$\leq \frac{2^{j-\alpha_j}}{\Gamma(j-\alpha_j+1)} \frac{C_2^*}{n^{\sigma-j-1}} \omega_{m-\sigma+1}\left(f^{(\sigma-1)}, \frac{1}{n}\right), \tag{5.13}$$

$j = h+1, ..., \sigma - 2.$

Set

$$l_j :\equiv \sup_{x\in[-1,1]} \left|\alpha_h^{-1}(x)\,\alpha_j(x)\right|, \quad h \le j \le k. \tag{5.14}$$

When $j = 1,...,h$ we derive

$$\max_{-1\le x\le 1} \left|\left(D_{*-1}^{\alpha_j} f\right)(x) - \left(D_{*-1}^{\alpha_j} Q_n\right)(x)\right| \le \frac{C_2^*}{n^{\sigma-k-1}} \omega_{m-\sigma+1}\left(f^{(\sigma-1)}, \frac{1}{n}\right)$$
$$\cdot \left[\left(\sum_{\tau=h}^{k} l_\tau \frac{2^{\tau-\alpha_\tau}}{\Gamma(\tau-\alpha_\tau+1)}\right)\left(\sum_{\lambda=0}^{h-j} \frac{2^{h-\alpha_j-\lambda}}{\lambda!\Gamma(h-\alpha_j-\lambda+1)}\right) + \frac{2^{j-\alpha_j}}{\Gamma(j-\alpha_j+1)}\right]. \tag{5.15}$$

Finally it holds

$$\sup_{-1\le x\le 1} |f(x) - Q_n(x)|$$
$$\le \frac{C_2^*}{n^{\sigma-k-1}} \omega_{m-\sigma+1}\left(f^{(\sigma-1)}, \frac{1}{n}\right)\left[\frac{1}{h!}\sum_{\tau=h}^{k} l_\tau \frac{2^{\tau-\alpha_\tau}}{\Gamma(\tau-\alpha_\tau+1)} + 1\right]. \tag{5.16}$$

Proof. Set $\alpha_0 = 0$, thus $\lceil\alpha_0\rceil = 0$. We have $\lceil\alpha_j\rceil = j$, $j = 1, ..., \sigma - 2$.

Let Q_n *as in Lemma* 5.3.

We notice that ($x \in [-1,1]$)

$$\left|\left(D_{*-1}^{\alpha_j} f\right)(x) - \left(D_{*-1}^{\alpha_j} Q_n\right)(x)\right|$$
$$= \frac{1}{\Gamma(j-\alpha_j)}\left|\int_{-1}^{x} (x-t)^{j-\alpha_j-1} f^{(j)}(t)\,dt - \int_{-1}^{x} (x-t)^{j-\alpha_j-1} Q_n^{(j)}(t)\,dt\right| \tag{5.17}$$

$$= \frac{1}{\Gamma(j-\alpha_j)}\left|\int_{-1}^{x} (x-t)^{j-\alpha_j-1}\left(f^{(j)}(t) - Q_n^{(j)}(t)\right)dt\right|$$

$$\le \frac{1}{\Gamma(j-\alpha_j)}\int_{-1}^{x} (x-t)^{j-\alpha_j-1}\left|f^{(j)}(t) - Q_n^{(j)}(t)\right|dt \tag{5.18}$$

$$\overset{(5.9)}{\le} \frac{1}{\Gamma(j-\alpha_j)}\left(\int_{-1}^{x} (x-t)^{j-\alpha_j-1}dt\right)\frac{C_2^*}{n^{\sigma-j-1}}\omega_{m-\sigma+1}\left(f^{(\sigma-1)}, \frac{1}{n}\right)$$

$$= \frac{1}{\Gamma(j-\alpha_j)}\frac{(x+1)^{j-\alpha_j}}{(j-\alpha_j)}\frac{C_2^*}{n^{\sigma-j-1}}\omega_{m-\sigma+1}\left(f^{(\sigma-1)}, \frac{1}{n}\right) \tag{5.19}$$

$$= \frac{(x+1)^{j-\alpha_j}}{\Gamma(j-\alpha_j+1)}\frac{C_2^*}{n^{\sigma-j-1}}\omega_{m-\sigma+1}\left(f^{(\sigma-1)}, \frac{1}{n}\right)$$

$$\le \frac{2^{j-\alpha_j}}{\Gamma(j-\alpha_j+1)}\frac{C_2^*}{n^{\sigma-j-1}}\omega_{m-\sigma+1}\left(f^{(\sigma-1)}, \frac{1}{n}\right). \tag{5.20}$$

Hence

$$\left\|D_{*-1}^{\alpha_j} f - D_{*-1}^{\alpha_j} Q_n\right\|_{\infty,[-1,1]} \leq \frac{2^{j-\alpha_j}}{\Gamma(j-\alpha_j+1)} \frac{C_2^*}{n^{\sigma-j-1}} \omega_{m-\sigma+1}\left(f^{(\sigma-1)}, \frac{1}{n}\right), \tag{5.21}$$

$j = 0, 1, ..., \sigma - 2$.

We set

$$\rho_n := C_2^* \omega_{m-\sigma+1}\left(f^{(\sigma-1)}, \frac{1}{n}\right) \left(\sum_{j=h}^{k} l_j \frac{2^{j-\alpha_j}}{\Gamma(j-\alpha_j+1) n^{\sigma-j-1}}\right). \tag{5.22}$$

I. Suppose, throughout $[0,1]$, $\alpha_h(x) \geq \alpha > 0$. Let $Q_n(x)$, $x \in [-1,1]$, the polynomial spline of order $m > 0$ with simple knots at the points x_{in}, $i = 1, ..., n-1$, on $[-1,1]$ ($x_{in} = -1 + i\frac{2}{n}$, $i = 0, 1, ..., n$, here $\overline{\Delta}_n = \frac{2}{n}$), so that

$$\max_{-1\leq x\leq 1} \left| D_{*-1}^{\alpha_j}\left(f(x) + \rho_n \frac{x^h}{h!}\right) - \left(D_{*-1}^{\alpha_j} Q_n\right)(x)\right|$$
$$\leq \frac{2^{j-\alpha_j}}{\Gamma(j-\alpha_j+1)} \frac{C_2^*}{n^{\sigma-j-1}} \omega_{m-\sigma+1}\left(f^{(\sigma-1)}, \frac{1}{n}\right), \tag{5.23}$$

$j = 0, 1, ..., \sigma - 2$.

When $j = h+1, ..., \sigma - 2$, then

$$\max_{-1\leq x\leq 1} \left|\left(D_{*-1}^{\alpha_j} f\right)(x) - \left(D_{*-1}^{\alpha_j} Q_n\right)(x)\right|$$
$$\leq \frac{2^{j-\alpha_j}}{\Gamma(j-\alpha_j+1)} \frac{C_2^*}{n^{\sigma-j-1}} \omega_{m-\sigma+1}\left(f^{(\sigma-1)}, \frac{1}{n}\right), \tag{5.24}$$

proving (5.13).

For $j = 1, ..., h$ we find that

$$D_{*-1}^{\alpha_j}\left(\frac{x^h}{h!}\right) = \sum_{\lambda=0}^{h-j} \frac{(-1)^\lambda (x+1)^{h-\alpha_j-\lambda}}{\lambda! \Gamma(h-\alpha_j-\lambda+1)}. \tag{5.25}$$

Therefore we get from (5.23)

$$\max_{-1\leq x\leq 1} \left|\left(D_{*-1}^{\alpha_j} f\right)(x) + \rho_n \left(\sum_{\lambda=0}^{h-j} \frac{(-1)^\lambda (x+1)^{h-\alpha_j-\lambda}}{\lambda! \Gamma(h-\alpha_j-\lambda+1)}\right) - \left(D_{*-1}^{\alpha_j} Q_n\right)(x)\right| \tag{5.26}$$

$$\leq \frac{2^{j-\alpha_j}}{\Gamma(j-\alpha_j+1)} \frac{C_2^*}{n^{\sigma-j-1}} \omega_{m-\sigma+1}\left(f^{(\sigma-1)}, \frac{1}{n}\right),$$

$j = 1, ..., h$.

Therefore we get for $j = 1, ..., h$, that

$$\max_{-1\leq x\leq 1} \left|\left(D_{*-1}^{\alpha_j} f\right)(x) - \left(D_{*-1}^{\alpha_j} Q_n\right)(x)\right| \tag{5.27}$$

$$\leq \rho_n \left(\sum_{\lambda=0}^{h-j} \frac{2^{h-\alpha_j-\lambda}}{\lambda! \Gamma(h-\alpha_j-\lambda+1)} \right) + \frac{2^{j-\alpha_j}}{\Gamma(j-\alpha_j+1)} \frac{C_2^*}{n^{\sigma-j-1}} \omega_{m-\sigma+1} \left(f^{(\sigma-1)}, \frac{1}{n} \right)$$

$$= C_2^* \omega_{m-\sigma+1} \left(f^{(\sigma-1)}, \frac{1}{n} \right) \left(\sum_{\bar{j}=h}^{k} l_{\bar{j}} \frac{2^{\bar{j}-\alpha_{\bar{j}}}}{\Gamma\left(\bar{j}-\alpha_{\bar{j}}+1\right) n^{\sigma-\bar{j}-1}} \right)$$

$$\cdot \left(\sum_{\lambda=0}^{h-j} \frac{2^{h-\alpha_j-\lambda}}{\lambda! \Gamma(h-\alpha_j-\lambda+1)} \right) + \frac{2^{j-\alpha_j}}{\Gamma(j-\alpha_j+1)} \frac{C_2^*}{n^{\sigma-j-1}} \omega_{m-\sigma+1} \left(f^{(\sigma-1)}, \frac{1}{n} \right)$$

$$= C_2^* \omega_{m-\sigma+1} \left(f^{(\sigma-1)}, \frac{1}{n} \right) \left[\left(\sum_{\bar{j}=h}^{k} l_{\bar{j}} \frac{2^{\bar{j}-\alpha_{\bar{j}}}}{\Gamma\left(\bar{j}-\alpha_{\bar{j}}+1\right)} \frac{1}{n^{\sigma-\bar{j}-1}} \right) \right. \tag{5.28}$$

$$\left. \cdot \left(\sum_{\lambda=0}^{h-j} \frac{2^{h-\alpha_j-\lambda}}{\lambda! \Gamma(h-\alpha_j-\lambda+1)} \right) + \frac{2^{j-\alpha_j}}{\Gamma(j-\alpha_j+1)} \frac{1}{n^{\sigma-j-1}} \right]$$

$$\leq C_2^* \omega_{m-\sigma+1} \left(f^{(\sigma-1)}, \frac{1}{n} \right) \frac{1}{n^{\sigma-k-1}} \left[\left(\sum_{\bar{j}=h}^{k} l_{\bar{j}} \frac{2^{\bar{j}-\alpha_{\bar{j}}}}{\Gamma\left(\bar{j}-\alpha_{\bar{j}}+1\right)} \right) \right. \tag{5.29}$$

$$\left. \cdot \left(\sum_{\lambda=0}^{h-j} \frac{2^{h-\alpha_j-\lambda}}{\lambda! \Gamma(h-\alpha_j-\lambda+1)} \right) + \frac{2^{j-\alpha_j}}{\Gamma(j-\alpha_j+1)} \right].$$

Hence for $j = 1, ..., h$ we derived (5.15):

$$\max_{-1 \leq x \leq 1} \left| \left(D_{*-1}^{\alpha_j} f \right)(x) - \left(D_{*-1}^{\alpha_j} Q_n \right)(x) \right| \leq \frac{C_2^*}{n^{\sigma-k-1}} \omega_{m-\sigma+1} \left(f^{(\sigma-1)}, \frac{1}{n} \right)$$
$$\cdot \left[\left(\sum_{\tau=h}^{k} l_\tau \frac{2^{\tau-\alpha_\tau}}{\Gamma(\tau-\alpha_\tau+1)} \right) \left(\sum_{\lambda=0}^{h-j} \frac{2^{h-\alpha_j-\lambda}}{\lambda! \Gamma(h-\alpha_j-\lambda+1)} \right) + \frac{2^{j-\alpha_j}}{\Gamma(j-\alpha_j+1)} \right]. \tag{5.30}$$

When $j = 0$ from (5.23) we obtain

$$\max_{-1 \leq x \leq 1} \left| f(x) + \rho_n \frac{x^h}{h!} - Q_n(x) \right| \leq \frac{C_2^*}{n^{\sigma-1}} \omega_{m-\sigma+1} \left(f^{(\sigma-1)}, \frac{1}{n} \right). \tag{5.31}$$

And

$$\max_{-1 \leq x \leq 1} |f(x) - Q_n(x)| \leq \frac{\rho_n}{h!} + \frac{C_2^*}{n^{\sigma-1}} \omega_{m-\sigma+1} \left(f^{(\sigma-1)}, \frac{1}{n} \right) \tag{5.32}$$

$$= \frac{C_2^*}{h!} \omega_{m-\sigma+1} \left(f^{(\sigma-1)}, \frac{1}{n} \right) \left(\sum_{\tau=h}^{k} l_\tau \frac{2^{\tau-\alpha_\tau}}{\Gamma(\tau-\alpha_\tau+1) n^{\sigma-\tau-1}} \right)$$

$$+\frac{C_2^*}{n^{\sigma-1}}\omega_{m-\sigma+1}\left(f^{(\sigma-1)},\frac{1}{n}\right)$$

$$=C_2^*\omega_{m-\sigma+1}\left(f^{(\sigma-1)},\frac{1}{n}\right)\left[\frac{1}{h!}\sum_{\tau=h}^{k}l_\tau\frac{2^{\tau-\alpha_\tau}}{\Gamma\left(\tau-\alpha_\tau+1\right)n^{\sigma-\tau-1}}+\frac{1}{n^{\sigma-1}}\right] \tag{5.33}$$

$$\leq\frac{C_2^*}{n^{\sigma-k-1}}\omega_{m-\sigma+1}\left(f^{(\sigma-1)},\frac{1}{n}\right)\left[\frac{1}{h!}\sum_{\tau=h}^{k}l_\tau\frac{2^{\tau-\alpha_\tau}}{\Gamma\left(\tau-\alpha_\tau+1\right)}+1\right].$$

Proving

$$\max_{-1\leq x\leq 1}\left|f\left(x\right)-Q_n\left(x\right)\right|$$

$$\leq\frac{C_2^*}{n^{\sigma-k-1}}\omega_{m-\sigma+1}\left(f^{(\sigma-1)},\frac{1}{n}\right)\left[\frac{1}{h!}\sum_{\tau=h}^{k}l_\tau\frac{2^{\tau-\alpha_\tau}}{\Gamma\left(\tau-\alpha_\tau+1\right)}+1\right], \tag{5.34}$$

So that (5.16) is established.

Also if $0\leq x\leq 1$, then

$$\alpha_h^{-1}\left(x\right)L^*\left(Q_n\left(x\right)\right)=\alpha_h^{-1}\left(x\right)L^*\left(f\left(x\right)\right)+\rho_n\frac{\left(x+1\right)^{h-\alpha_h}}{\Gamma\left(h-\alpha_h+1\right)} \tag{5.35}$$

$$+\sum_{j=h}^{k}\alpha_h^{-1}\left(x\right)\alpha_j\left(x\right)\left[D_{*-1}^{\alpha_j}Q_n\left(x\right)-D_{*-1}^{\alpha_j}f\left(x\right)-\frac{\rho_n}{h!}D_{*-1}^{\alpha_j}x^h\right]$$

$$\overset{(5.23)}{\geq}\rho_n\frac{(x+1)^{h-\alpha_h}}{\Gamma(h-\alpha_h+1)}-\left(\sum_{j=h}^{k}l_j\frac{2^{j-\alpha_j}}{\Gamma(j-\alpha_j+1)}\frac{C_2^*}{n^{\sigma-j-1}}\omega_{m-\sigma+1}\left(f^{(\sigma-1)},\frac{1}{n}\right)\right)$$

$$=\rho_n\frac{\left(x+1\right)^{h-\alpha_h}}{\Gamma\left(h-\alpha_h+1\right)}-\rho_n=\rho_n\left[\frac{\left(x+1\right)^{h-\alpha_h}}{\Gamma\left(h-\alpha_h+1\right)}-1\right]$$

$$=\rho_n\left[\frac{\left(x+1\right)^{h-\alpha_h}-\Gamma\left(h-\alpha_h+1\right)}{\Gamma\left(h-\alpha_h+1\right)}\right]\geq\rho_n\left[\frac{1-\Gamma\left(h-\alpha_h+1\right)}{\Gamma\left(h-\alpha_h+1\right)}\right]\geq 0. \tag{5.36}$$

Explanation: We know that $\Gamma\left(1\right)=1$, $\Gamma\left(2\right)=1$, and Γ is convex and positive on $(0,\infty)$. Here $0\leq h-\alpha_h<1$ and $1\leq h-\alpha_h+1<2$. Thus $\Gamma\left(h-\alpha_h+1\right)\leq 1$ and $1-\Gamma\left(h-\alpha_h+1\right)\geq 0$. Hence $L^*\left(Q_n\left(x\right)\right)\geq 0$, $x\in\left[0,1\right]$.

II. Suppose on $[0,1]$ that $\alpha_h\left(x\right)\leq\beta<0$. Let $Q_n\left(x\right)$, $x\in[-1,1]$, be the polynomial spline of order $m>0$, (as before), so that

$$\max_{-1\leq x\leq 1}\left|D_{*-1}^{\alpha_j}\left(f\left(x\right)-\rho_n\frac{x^h}{h!}\right)-\left(D_{*-1}^{\alpha_j}Q_n\right)\left(x\right)\right|$$

$$\leq \frac{2^{j-\alpha_j}}{\Gamma(j-\alpha_j+1)} \frac{C_2^*}{n^{\sigma-j-1}} \omega_{m-\sigma+1}\left(f^{(\sigma-1)}, \frac{1}{n}\right), \tag{5.37}$$

$j = 0, 1, ..., \sigma - 2$.

Similarly as before we obtain again inequalities of convergence (5.13), (5.15) and (5.16).

Also if $0 \leq x \leq 1$, then

$$\alpha_h^{-1}(x) L^*(Q_n(x)) = \alpha_h^{-1}(x) L^*(f(x)) - \rho_n \frac{(x+1)^{h-\alpha_h}}{\Gamma(h-\alpha_h+1)} \tag{5.38}$$

$$+\sum_{j=h}^{k} \alpha_h^{-1}(x) \alpha_j(x) \left[D_{*-1}^{\alpha_j} Q_n(x) - D_{*-1}^{\alpha_j} f(x) + \frac{\rho_n}{h!}\left(D_{*-1}^{\alpha_j} x^h\right)\right]$$

$$\overset{(5.37)}{\leq} -\rho_n \frac{(x+1)^{h-\alpha_h}}{\Gamma(h-\alpha_h+1)} + \sum_{j=h}^{k} l_j \frac{2^{j-\alpha_j}}{\Gamma(j-\alpha_j+1)} \frac{C_2^*}{n^{\sigma-j-1}} \omega_{m-\sigma+1}\left(f^{(\sigma-1)}, \frac{1}{n}\right) \tag{5.39}$$

$$= \rho_n \left(1 - \frac{(x+1)^{h-\alpha_h}}{\Gamma(h-\alpha_h+1)}\right) = \rho_n \left(\frac{\Gamma(h-\alpha_h+1) - (x+1)^{h-\alpha_h}}{\Gamma(h-\alpha_h+1)}\right) \tag{5.40}$$

$$\leq \rho_n \left(\frac{1-(x+1)^{h-\alpha_h}}{\Gamma(h-\alpha_h+1)}\right) \leq 0,$$

and hence again $L^*(Q_n(x)) \geq 0$, $x \in [0,1]$. □

Bibliography

1. G.A. Anastassiou, *Spline monotone approximation with linear differential operators*, Approx. Theory Appl. 5 (4) (1989), 61-67.
2. G.A. Anastassiou, *Spline left fractional monotone approximation involving left fractional differential operators*, CUBO, accepted 2014.
3. R. De Vore, G. Lorentz, *Constructive Approximation*, Springer-Verlag, Heidelberg, New York, 1993.
4. K. Diethelm, *The Analysis of Fractional Differential Equations*, Lecture Notes in Mathematics, Vol. 2004, 1st edition, Springer, New York, Heidelberg, 2010.
5. L.L. Schumaker, *Spline Functions: Basic Theory*, John Wiley and Sons, Inc., New York, 1981.

Chapter 6

Spline Right Fractional Monotone Approximation Theory Using Right Fractional Differential Operators

Let $f \in C^s([-1,1])$, $s \in \mathbb{N}$ and L^* be a linear right fractional differential operator such that $L^*(f) \geq 0$ on $[-1,0]$. Then there exists a sequence Q_n, $n \in \mathbb{N}$ of polynomial splines with equally spaced knots of given fixed order such that $L^*(Q_n) \geq 0$ on $[-1,0]$. Furthermore f is approximated with rates right fractionally and simultaneously by Q_n in the uniform norm. This constrained right fractional approximation on $[-1,1]$ is given via inequalities involving a higher modulus of smoothness of $f^{(s)}$. It follows [2].

6.1 Introduction

Let $[a,b] \subset \mathbb{R}$ and for $n \geq 1$ consider the partition Δ_n with points $x_{in} = a + i\left(\frac{b-a}{n}\right)$, $i = 0,1,...,n$. Hence $\overline{\Delta}_n \equiv \max_{1 \leq i \leq n}(x_{in} - x_{i-1,n}) = \frac{b-a}{n}$.

Let $S_m(\Delta_n)$ be the space of polynomial splines of order $m > 0$ with simple knots at the points x_{in}, $i = 1,...,n-1$. Then there exists a linear operator Q_n : $Q_n \equiv Q_n(f)$, mapping $B[a,b]$: the space of bounded real valued functions f on $[a,b]$, into $S_m(\Delta_n)$ (see [5], p. 224, Theorem 6.18).

From the same reference [5], p. 227, Corollary 6.21, we get

Corollary 6.1. *Let $1 \leq \sigma \leq m$, $n \geq 1$. Then for all $f \in C^{\sigma-1}[a,b]$; $r = 0,...,\sigma-1$,*

$$\left\| f^{(r)} - Q_n^{(r)} \right\|_\infty \leq C_1 \left(\frac{b-a}{n}\right)^{\sigma-r-1} \omega_{m-\sigma+1}\left(f^{(\sigma-1)}, \frac{b-a}{n}\right), \tag{6.1}$$

where C_1 depends only on m, $C_1 = C_1(m)$.

By denoting $C_2 = C_1 \max_{0 \leq r \leq \sigma-1}(b-a)^{\sigma-r-1}$ we obtain

Lemma 6.1. *([1]) Let $1 \leq \sigma \leq m$, $n \geq 1$. Then for all $f \in C^{\sigma-1}[a,b]$; $r = 0,...,\sigma-1$,*

$$\left\| f^{(r)} - Q_n^{(r)} \right\|_\infty \leq \frac{C_2}{n^{\sigma-r-1}} \omega_{m-\sigma+1}\left(f^{(\sigma-1)}, \frac{b-a}{n}\right), \tag{6.2}$$

where C_2 depends only on m, σ and $b-a$. Here $\omega_{m-\sigma+1}$ is the usual modulus of smoothness of order $m-\sigma+1$.

We are motivated by

Theorem 6.1. *([1]) Let h, k, σ, m be integers, $0 \leq h \leq k \leq \sigma - 1$, $\sigma \leq m$ and let $f \in C^{\sigma-1}[a,b]$. Let $\alpha_j(x) \in B[a,b]$, $j = h, h+1, ..., k$ and suppose that $\alpha_h(x) \geq \alpha > 0$ or $\alpha_h(x) \leq \beta < 0$ for all $x \in [a,b]$. Take the linear differential operator*

$$L = \sum_{j=h}^{k} \alpha_j(x) \left[\frac{d^j}{dx^j}\right] \tag{6.3}$$

and assume, throughout $[a,b]$,

$$L(f) \geq 0. \tag{6.4}$$

Then, for every integer $n \geq 1$, there is a polynomial spline function $Q_n(x)$ of order m with simple knots at $\left\{a + i\left(\frac{b-a}{n}\right),\ i = 1, ..., n-1\right\}$ such that $L(Q_n) \geq 0$ throughout $[a,b]$ and

$$\left\|f^{(r)} - Q_n^{(r)}\right\|_\infty \leq \frac{C}{n^{\sigma-k-1}} \omega_{m-\sigma+1}\left(f^{(\sigma-1)}, \frac{b-a}{n}\right), \quad 0 \leq r \leq h. \tag{6.5}$$

Moreover, we find

$$\left\|f^{(r)} - Q_n^{(r)}\right\|_\infty \leq \frac{C}{n^{\sigma-r-1}} \omega_{m-\sigma+1}\left(f^{(\sigma-1)}, \frac{b-a}{n}\right), \quad h+1 \leq r \leq \sigma - 1, \tag{6.6}$$

where C is a constant independent of f and n. It depends only on m, σ, L, a, b.

Next we specialize on the case of $a = -1$, $b = 1$. That is working on $[-1,1]$. By Lemma 6.1 we get

Lemma 6.2. *Let $1 \leq \sigma \leq m$, $n \geq 1$. Then for all $f \in C^{\sigma-1}([-1,1])$; $j = 0, 1, ..., \sigma - 1$,*

$$\left\|f^{(j)} - Q_n^{(j)}\right\|_\infty \leq \frac{C_2}{n^{\sigma-j-1}} \omega_{m-\sigma+1}\left(f^{(\sigma-1)}, \frac{2}{n}\right), \tag{6.7}$$

where $C_2 := C_2(m,\sigma) := C_1(m) 2^{\sigma-1}$.

Since

$$\omega_{m-\sigma+1}\left(f^{(\sigma-1)}, \frac{2}{n}\right) \leq 2^{m-\sigma+1} \omega_{m-\sigma+1}\left(f^{(\sigma-1)}, \frac{1}{n}\right) \tag{6.8}$$

(see [3], p. 45), we get

Lemma 6.3. *Let $1 \leq \sigma \leq m$, $n \geq 1$. Then for all $f \in C^{\sigma-1}([-1,1])$; $j = 0, 1, ..., \sigma - 1$,*

$$\left\|f^{(j)} - Q_n^{(j)}\right\|_\infty \leq \frac{C_2^*}{n^{\sigma-j-1}} \omega_{m-\sigma+1}\left(f^{(\sigma-1)}, \frac{1}{n}\right), \tag{6.9}$$

where $C_2^* := C_2^*(m,\sigma) := C_1(m) 2^m$.

We use a lot in this chapter Lemma 6.3.

In this chapter we extend Theorem 6.1 over $[-1,1]$ to the right fractional level. Indeed here L is replaced by L^*, a linear right Caputo fractional differential operator. Now the monotonicity property is only true on the critical interval $[-1,0]$. Simultaneous fractional convergence remains true on all of $[-1,1]$.

We make

Definition 6.1. ([4]) Let $\alpha > 0$ and $\lceil \alpha \rceil = m$, ($\lceil \cdot \rceil$ ceiling of the number). Consider $f \in C^m([-1,1])$. We define the right Caputo fractional derivative of f of order α as follows:

$$\left(D_{1-}^{\alpha} f\right)(x) = \frac{(-1)^m}{\Gamma(m-\alpha)} \int_x^1 (t-x)^{m-\alpha-1} f^{(m)}(t)\, dt, \tag{6.10}$$

for any $x \in [-1,1]$, where Γ is the gamma function.

We set

$$D_{1-}^{0} f(x) := f(x),$$

$$D_{1-}^{m} f(x) := (-1)^m f^{(m)}(x), \quad \forall\, x \in [-1,1]. \tag{6.11}$$

6.2 Main Result

We give

Theorem 6.2. *Let h, k, σ, m be integers, $1 \leq \sigma \leq m$, $n \in \mathbb{N}$, h is even, with $0 \leq h \leq k \leq \sigma - 2$ and let $f \in C^{\sigma-1}([-1,1])$, with $f^{(\sigma-1)}$ having modulus of smoothness $\omega_{m-\sigma+1}\left(f^{(\sigma-1)}, \delta\right)$ there, $\delta > 0$. Let $\alpha_j(x)$, $j = h, h+1, ..., k$ be real functions, defined and bounded on $[-1,1]$ and suppose $\alpha_h(x)$ is either $\geq \alpha > 0$ or $\leq \beta < 0$ on $[-1,0]$. Let the real numbers $\alpha_0 = 0 < \alpha_1 < 1 < \alpha_2 < 2 < ... < \alpha_{\sigma-2} < \sigma - 2$. Here $D_{1-}^{\alpha_j} f$ stands for the right Caputo fractional derivative of f of order α_j anchored at 1. Consider the linear right fractional differential operator*

$$L^* := \sum_{j=h}^{k} \alpha_j(x) \left[D_{1-}^{\alpha_j}\right] \tag{6.12}$$

and suppose, throughout $[-1,0]$, $L^*(f) \geq 0$.

Then, for every integer $n \geq 1$, there exists a polynomial spline function $Q_n(x)$ of order $m > 0$ with simple knots at $\left\{-1 + i\frac{2}{n},\ i = 1, ..., n-1\right\}$ such that $L^*(Q_n) \geq 0$ throughout $[-1,0]$, and

$$\sup_{-1 \leq x \leq 1} \left| \left(D_{1-}^{\alpha_j} f\right)(x) - \left(D_{1-}^{\alpha_j} Q_n\right)(x) \right|$$

$$\leq \frac{2^{j-\alpha_j}}{\Gamma(j-\alpha_j+1)} \frac{C_2^*}{n^{\sigma-j-1}} \omega_{m-\sigma+1}\left(f^{(\sigma-1)}, \frac{1}{n}\right), \tag{6.13}$$

$j = h+1, ..., \sigma - 2.$

Set

$$l_j :\equiv \sup_{x \in [-1,1]} \left| \alpha_h^{-1}(x) \alpha_j(x) \right|, \quad h \leq j \leq k. \tag{6.14}$$

When $j = 1, ..., h$ *we derive*

$$\sup_{-1 \leq x \leq 1} \left| \left(D_{1-}^{\alpha_j} f \right)(x) - \left(D_{1-}^{\alpha_j} Q_n \right)(x) \right| \leq \frac{C_2^*}{n^{\sigma-k-1}} \omega_{m-\sigma+1} \left(f^{(\sigma-1)}, \frac{1}{n} \right)$$
$$\cdot \left[\left(\sum_{\tau=h}^{k} l_\tau \frac{2^{\tau-\alpha_\tau}}{\Gamma(\tau - \alpha_\tau + 1)} \right) \left(\sum_{\lambda=0}^{h-j} \frac{2^{h-\alpha_j-\lambda}}{\lambda! \Gamma(h - \alpha_j - \lambda + 1)} \right) + \frac{2^{j-\alpha_j}}{\Gamma(j - \alpha_j + 1)} \right]. \tag{6.15}$$

Finally it holds

$$\sup_{-1 \leq x \leq 1} |f(x) - Q_n(x)|$$
$$\leq \frac{C_2^*}{n^{\sigma-k-1}} \omega_{m-\sigma+1} \left(f^{(\sigma-1)}, \frac{1}{n} \right) \left[\frac{1}{h!} \sum_{\tau=h}^{k} l_\tau \frac{2^{\tau-\alpha_\tau}}{\Gamma(\tau - \alpha_\tau + 1)} + 1 \right]. \tag{6.16}$$

Proof. Set $\alpha_0 = 0$, thus $\lceil \alpha_0 \rceil = 0$. We have $\lceil \alpha_j \rceil = j$, $j = 1, ..., \sigma - 2$.
Let Q_n *as in Lemma* 6.3.
We notice that ($x \in [-1, 1]$)

$$\left| \left(D_{1-}^{\alpha_j} f \right)(x) - \left(D_{1-}^{\alpha_j} Q_n \right)(x) \right|$$
$$= \frac{1}{\Gamma(j - \alpha_j)} \left| \int_x^1 (t - x)^{j-\alpha_j-1} f^{(j)}(t)\, dt - \int_x^1 (t - x)^{j-\alpha_j-1} Q_n^{(j)}(t)\, dt \right| \tag{6.17}$$

$$= \frac{1}{\Gamma(j - \alpha_j)} \left| \int_x^1 (t - x)^{j-\alpha_j-1} \left(f^{(j)}(t) - Q_n^{(j)}(t) \right) dt \right|$$

$$\leq \frac{1}{\Gamma(j - \alpha_j)} \int_x^1 (t - x)^{j-\alpha_j-1} \left| f^{(j)}(t) - Q_n^{(j)}(t) \right| dt \tag{6.18}$$

$$\overset{(6.9)}{\leq} \frac{1}{\Gamma(j - \alpha_j)} \left(\int_x^1 (t - x)^{j-\alpha_j-1} dt \right) \frac{C_2^*}{n^{\sigma-j-1}} \omega_{m-\sigma+1} \left(f^{(\sigma-1)}, \frac{1}{n} \right)$$

$$= \frac{1}{\Gamma(j - \alpha_j)} \frac{(1 - x)^{j-\alpha_j}}{(j - \alpha_j)} \frac{C_2^*}{n^{\sigma-j-1}} \omega_{m-\sigma+1} \left(f^{(\sigma-1)}, \frac{1}{n} \right) \tag{6.19}$$

$$= \frac{(1 - x)^{j-\alpha_j}}{\Gamma(j - \alpha_j + 1)} \frac{C_2^*}{n^{\sigma-j-1}} \omega_{m-\sigma+1} \left(f^{(\sigma-1)}, \frac{1}{n} \right)$$

$$\leq \frac{2^{j-\alpha_j}}{\Gamma(j - \alpha_j + 1)} \frac{C_2^*}{n^{\sigma-j-1}} \omega_{m-\sigma+1} \left(f^{(\sigma-1)}, \frac{1}{n} \right). \tag{6.20}$$

Hence

$$\left\|D_{1-}^{\alpha_j} f - D_{1-}^{\alpha_j} Q_n\right\|_{\infty,[-1,1]} \leq \frac{2^{j-\alpha_j}}{\Gamma(j-\alpha_j+1)} \frac{C_2^*}{n^{\sigma-j-1}} \omega_{m-\sigma+1}\left(f^{(\sigma-1)}, \frac{1}{n}\right), \tag{6.21}$$

$j = 0, 1, ..., \sigma - 2$.

We set

$$\rho_n := C_2^* \omega_{m-\sigma+1}\left(f^{(\sigma-1)}, \frac{1}{n}\right) \left(\sum_{j=h}^{k} l_j \frac{2^{j-\alpha_j}}{\Gamma(j-\alpha_j+1) n^{\sigma-j-1}}\right). \tag{6.22}$$

I. Suppose, throughout $[-1, 0]$, $\alpha_h(x) \geq \alpha > 0$. Let $Q_n(x)$, $x \in [-1, 1]$, the polynomial spline of order $m > 0$ with simple knots at the points x_{in}, $i = 1, ..., n-1$, on $[-1, 1]$ ($x_{in} = -1 + i\frac{2}{n}$, $i = 0, 1, ..., n$, here $\overline{\Delta}_n = \frac{2}{n}$), so that

$$\max_{-1 \leq x \leq 1} \left| D_{1-}^{\alpha_j}\left(f(x) + \rho_n \frac{x^h}{h!}\right) - \left(D_{1-}^{\alpha_j} Q_n\right)(x)\right|$$

$$\leq \frac{2^{j-\alpha_j}}{\Gamma(j-\alpha_j+1)} \frac{C_2^*}{n^{\sigma-j-1}} \omega_{m-\sigma+1}\left(f^{(\sigma-1)}, \frac{1}{n}\right), \tag{6.23}$$

$j = 0, 1, ..., \sigma - 2$.

When $j = h+1, ..., \sigma - 2$, then

$$\max_{-1 \leq x \leq 1} \left|\left(D_{1-}^{\alpha_j} f\right)(x) - \left(D_{1-}^{\alpha_j} Q_n\right)(x)\right|$$

$$\leq \frac{2^{j-\alpha_j}}{\Gamma(j-\alpha_j+1)} \frac{C_2^*}{n^{\sigma-j-1}} \omega_{m-\sigma+1}\left(f^{(\sigma-1)}, \frac{1}{n}\right), \tag{6.24}$$

proving (6.13).

For $j = 1, ..., h$ we find that

$$D_{1-}^{\alpha_j}\left(\frac{x^h}{h!}\right) = (-1)^h \sum_{\lambda=0}^{h-j} \frac{(-1)^\lambda (1-x)^{h-\alpha_j-\lambda}}{\lambda! \Gamma(h-\alpha_j-\lambda+1)}. \tag{6.25}$$

Therefore we get from (6.23)

$$\max_{-1 \leq x \leq 1} \left|\left(D_{1-}^{\alpha_j} f\right)(x) + \rho_n \left((-1)^h \sum_{\lambda=0}^{h-j} \frac{(-1)^\lambda (1-x)^{h-\alpha_j-\lambda}}{\lambda! \Gamma(h-\alpha_j-\lambda+1)}\right) - \left(D_{1-}^{\alpha_j} Q_n\right)(x)\right| \tag{6.26}$$

$$\leq \frac{2^{j-\alpha_j}}{\Gamma(j-\alpha_j+1)} \frac{C_2^*}{n^{\sigma-j-1}} \omega_{m-\sigma+1}\left(f^{(\sigma-1)}, \frac{1}{n}\right),$$

$j = 1, ..., h$.

Therefore we get for $j = 1, ..., h$, that

$$\max_{-1 \leq x \leq 1} \left|\left(D_{1-}^{\alpha_j} f\right)(x) - \left(D_{1-}^{\alpha_j} Q_n\right)(x)\right| \tag{6.27}$$

$$\leq \rho_n \left(\sum_{\lambda=0}^{h-j} \frac{2^{h-\alpha_j-\lambda}}{\lambda! \Gamma(h-\alpha_j-\lambda+1)}\right) + \frac{2^{j-\alpha_j}}{\Gamma(j-\alpha_j+1)} \frac{C_2^*}{n^{\sigma-j-1}} \omega_{m-\sigma+1}\left(f^{(\sigma-1)}, \frac{1}{n}\right)$$

$$= C_2^* \omega_{m-\sigma+1}\left(f^{(\sigma-1)}, \frac{1}{n}\right)\left(\sum_{\bar{j}=h}^{k} l_{\bar{j}} \frac{2^{\bar{j}-\alpha_{\bar{j}}}}{\Gamma\left(\bar{j}-\alpha_{\bar{j}}+1\right) n^{\sigma-\bar{j}-1}}\right)$$

$$\cdot\left(\sum_{\lambda=0}^{h-j} \frac{2^{h-\alpha_j-\lambda}}{\lambda!\Gamma\left(h-\alpha_j-\lambda+1\right)}\right)+\frac{2^{j-\alpha_j}}{\Gamma\left(j-\alpha_j+1\right)} \frac{C_2^*}{n^{\sigma-j-1}} \omega_{m-\sigma+1}\left(f^{(\sigma-1)}, \frac{1}{n}\right)$$

$$= C_2^* \omega_{m-\sigma+1}\left(f^{(\sigma-1)}, \frac{1}{n}\right)\left[\left(\sum_{\bar{j}=h}^{k} l_{\bar{j}} \frac{2^{\bar{j}-\alpha_{\bar{j}}}}{\Gamma\left(\bar{j}-\alpha_{\bar{j}}+1\right)} \frac{1}{n^{\sigma-\bar{j}-1}}\right)\right. \tag{6.28}$$

$$\left.\cdot\left(\sum_{\lambda=0}^{h-j} \frac{2^{h-\alpha_j-\lambda}}{\lambda!\Gamma\left(h-\alpha_j-\lambda+1\right)}\right)+\frac{2^{j-\alpha_j}}{\Gamma\left(j-\alpha_j+1\right)} \frac{1}{n^{\sigma-j-1}}\right]$$

$$\leq C_2^* \omega_{m-\sigma+1}\left(f^{(\sigma-1)}, \frac{1}{n}\right) \frac{1}{n^{\sigma-k-1}}\left[\left(\sum_{\bar{j}=h}^{k} l_{\bar{j}} \frac{2^{\bar{j}-\alpha_{\bar{j}}}}{\Gamma\left(\bar{j}-\alpha_{\bar{j}}+1\right)}\right)\right. \tag{6.29}$$

$$\left.\cdot\left(\sum_{\lambda=0}^{h-j} \frac{2^{h-\alpha_j-\lambda}}{\lambda!\Gamma\left(h-\alpha_j-\lambda+1\right)}\right)+\frac{2^{j-\alpha_j}}{\Gamma\left(j-\alpha_j+1\right)}\right].$$

Hence for $j=1,...,h$ we derived (6.15):

$$\max_{-1 \leq x \leq 1}\left|\left(D_{1-}^{\alpha_j} f\right)(x)-\left(D_{1-}^{\alpha_j} Q_n\right)(x)\right| \leq \frac{C_2^*}{n^{\sigma-k-1}} \omega_{m-\sigma+1}\left(f^{(\sigma-1)}, \frac{1}{n}\right)$$

$$\cdot\left[\left(\sum_{\tau=h}^{k} l_\tau \frac{2^{\tau-\alpha_\tau}}{\Gamma\left(\tau-\alpha_\tau+1\right)}\right)\left(\sum_{\lambda=0}^{h-j} \frac{2^{h-\alpha_j-\lambda}}{\lambda!\Gamma\left(h-\alpha_j-\lambda+1\right)}\right)+\frac{2^{j-\alpha_j}}{\Gamma\left(j-\alpha_j+1\right)}\right]. \tag{6.30}$$

When $j=0$ from (6.23) we obtain

$$\max_{-1 \leq x \leq 1}\left|f(x)+\rho_n \frac{x^h}{h!}-Q_n(x)\right| \leq \frac{C_2^*}{n^{\sigma-1}} \omega_{m-\sigma+1}\left(f^{(\sigma-1)}, \frac{1}{n}\right). \tag{6.31}$$

And

$$\max_{-1 \leq x \leq 1}\left|f(x)-Q_n(x)\right| \leq \frac{\rho_n}{h!}+\frac{C_2^*}{n^{\sigma-1}} \omega_{m-\sigma+1}\left(f^{(\sigma-1)}, \frac{1}{n}\right) \tag{6.32}$$

$$= \frac{C_2^*}{h!} \omega_{m-\sigma+1}\left(f^{(\sigma-1)}, \frac{1}{n}\right)\left(\sum_{\tau=h}^{k} l_\tau \frac{2^{\tau-\alpha_\tau}}{\Gamma\left(\tau-\alpha_\tau+1\right) n^{\sigma-\tau-1}}\right)$$

$$+\frac{C_2^*}{n^{\sigma-1}} \omega_{m-\sigma+1}\left(f^{(\sigma-1)}, \frac{1}{n}\right)$$

$$= C_2^* \omega_{m-\sigma+1}\left(f^{(\sigma-1)}, \frac{1}{n}\right)\left[\frac{1}{h!}\sum_{\tau=h}^{k} l_\tau \frac{2^{\tau-\alpha_\tau}}{\Gamma(\tau-\alpha_\tau+1)\, n^{\sigma-\tau-1}} + \frac{1}{n^{\sigma-1}}\right] \tag{6.33}$$

$$\leq \frac{C_2^*}{n^{\sigma-k-1}} \omega_{m-\sigma+1}\left(f^{(\sigma-1)}, \frac{1}{n}\right)\left[\frac{1}{h!}\sum_{\tau=h}^{k} l_\tau \frac{2^{\tau-\alpha_\tau}}{\Gamma(\tau-\alpha_\tau+1)} + 1\right].$$

Proving

$$\max_{-1\leq x\leq 1} |f(x) - Q_n(x)|$$

$$\leq \frac{C_2^*}{n^{\sigma-k-1}} \omega_{m-\sigma+1}\left(f^{(\sigma-1)}, \frac{1}{n}\right)\left[\frac{1}{h!}\sum_{\tau=h}^{k} l_\tau \frac{2^{\tau-\alpha_\tau}}{\Gamma(\tau-\alpha_\tau+1)} + 1\right], \tag{6.34}$$

So that (6.16) is established.

Also if $-1 \leq x \leq 0$, then

$$\alpha_h^{-1}(x) L^*(Q_n(x)) = \alpha_h^{-1}(x) L^*(f(x)) + \rho_n \frac{(1-x)^{h-\alpha_h}}{\Gamma(h-\alpha_h+1)} \tag{6.35}$$

$$+\sum_{j=h}^{k} \alpha_h^{-1}(x)\alpha_j(x)\left[D_{1-}^{\alpha_j} Q_n(x) - D_{1-}^{\alpha_j} f(x) - \frac{\rho_n}{h!} D_{1-}^{\alpha_j} x^h\right]$$

$$\overset{(6.23)}{\geq} \rho_n \frac{(1-x)^{h-\alpha_h}}{\Gamma(h-\alpha_h+1)} - \left(\sum_{j=h}^{k} l_j \frac{2^{j-\alpha_j}}{\Gamma(j-\alpha_j+1)} \frac{C_2^*}{n^{\sigma-j-1}} \omega_{m-\sigma+1}\left(f^{(\sigma-1)}, \frac{1}{n}\right)\right)$$

$$= \rho_n \frac{(1-x)^{h-\alpha_h}}{\Gamma(h-\alpha_h+1)} - \rho_n = \rho_n\left[\frac{(1-x)^{h-\alpha_h}}{\Gamma(h-\alpha_h+1)} - 1\right]$$

$$= \rho_n\left[\frac{(1-x)^{h-\alpha_h} - \Gamma(h-\alpha_h+1)}{\Gamma(h-\alpha_h+1)}\right] \geq \rho_n\left[\frac{1-\Gamma(h-\alpha_h+1)}{\Gamma(h-\alpha_h+1)}\right] \geq 0. \tag{6.36}$$

Explanation: We know that $\Gamma(1) = 1$, $\Gamma(2) = 1$, and Γ is convex and positive on $(0,\infty)$. Here $0 \leq h-\alpha_h < 1$ and $1 \leq h-\alpha_h+1 < 2$. Thus $\Gamma(h-\alpha_h+1) \leq 1$ and $1-\Gamma(h-\alpha_h+1) \geq 0$. Hence $L^*(Q_n(x)) \geq 0$, $x \in [-1,0]$.

II. Suppose on $[-1,0]$ that $\alpha_h(x) \leq \beta < 0$. Let $Q_n(x)$, $x \in [-1,1]$, be the polynomial spline of order $m > 0$, (as before), so that

$$\max_{-1\leq x\leq 1}\left|D_{1-}^{\alpha_j}\left(f(x) - \rho_n \frac{x^h}{h!}\right) - \left(D_{1-}^{\alpha_j} Q_n\right)(x)\right|$$

$$\leq \frac{2^{j-\alpha_j}}{\Gamma(j-\alpha_j+1)} \frac{C_2^*}{n^{\sigma-j-1}} \omega_{m-\sigma+1}\left(f^{(\sigma-1)}, \frac{1}{n}\right), \tag{6.37}$$

$j = 0, 1, ..., \sigma-2.$

Similarly as before we obtain again inequalities of convergence (6.13), (6.15) and (6.16).

Also if $-1 \le x \le 0$, then

$$\alpha_h^{-1}(x) L^*(Q_n(x)) = \alpha_h^{-1}(x) L^*(f(x)) - \rho_n \frac{(1-x)^{h-\alpha_h}}{\Gamma(h-\alpha_h+1)} \tag{6.38}$$

$$+\sum_{j=h}^{k} \alpha_h^{-1}(x)\alpha_j(x)\left[D_{1-}^{\alpha_j} Q_n(x) - D_{1-}^{\alpha_j} f(x) + \frac{\rho_n}{h!}\left(D_{1-}^{\alpha_j} x^h\right)\right]$$

$$\overset{(6.37)}{\le} -\rho_n \frac{(1-x)^{h-\alpha_h}}{\Gamma(h-\alpha_h+1)} + \sum_{j=h}^{k} l_j \frac{2^{j-\alpha_j}}{\Gamma(j-\alpha_j+1)} \frac{C_2^*}{n^{\sigma-j-1}} \omega_{m-\sigma+1}\left(f^{(\sigma-1)}, \frac{1}{n}\right) \tag{6.39}$$

$$= \rho_n\left(1 - \frac{(1-x)^{h-\alpha_h}}{\Gamma(h-\alpha_h+1)}\right) = \rho_n\left(\frac{\Gamma(h-\alpha_h+1) - (1-x)^{h-\alpha_h}}{\Gamma(h-\alpha_h+1)}\right) \tag{6.40}$$

$$\le \rho_n\left(\frac{1-(1-x)^{h-\alpha_h}}{\Gamma(h-\alpha_h+1)}\right) \le 0,$$

and hence again $L^*(Q_n(x)) \ge 0$, $x \in [-1, 0]$. □

Bibliography

1. G.A. Anastassiou, *Spline monotone approximation with linear differential operators*, Approx. Theory Appl. 5 (4) (1989), 61-67.
2. G.A. Anastassiou, *Spline right fractional monotone approximation involving right fractional differential operators*, J. Concrete Applicable Math. 13 (1-2) (2015), 76-84.
3. R. De Vore, G. Lorentz, *Constructive Approximation*, Springer-Verlag, Heidelberg, New York, 1993.
4. A.M.A. El-Sayed, M. Gaber, *On the finite Caputo and finite Riesz derivatives*, Electron. J. Theoret. Phys. 3 (12) (2006), 81-95.
5. L.L. Schumaker, *Spline Functions: Basic Theory*, John Wiley and Sons, Inc., New York, 1981.

Chapter 7

Complete Fractional Monotone Approximation Theory

Here is developed the theory of complete fractional simultaneous monotone uniform polynomial approximation with rates using mixed fractional linear differential operators.

To achieve that, we establish first ordinary simultaneous polynomial approximation with respect to the highest order right and left fractional derivatives of the function under approximation using their moduli of continuity. Then we derive the complete right and left fractional simultaneous polynomial approximation with rates, as well we treat their affine combination. Based on the last and elegant analytical techniques, we derive preservation of monotonicity by mixed fractional linear differential operators. We study special cases. It follows [3].

7.1 Introduction

The topic of monotone approximation started in [6] has become a major trend in approximation theory. A typical problem in this subject is: given a positive integer k, approximate a given function whose kth derivative is ≥ 0 by polynomials having this property.

In [2] the authors replaced the kth derivative with a linear differential operator of order k. We mention this motivating result.

Theorem 7.1. *Let h, k, p be integers, $0 \leq h \leq k \leq p$ and let f be a real function, $f^{(p)}$ continuous in $[-1, 1]$ with first modulus of continuity $\omega_1\left(f^{(p)}, x\right)$ there. Let $a_j(x)$, $j = h, h+1, ..., k$ be real functions, defined and bounded on $[-1, 1]$ and assume $a_h(x)$ is either $\geq$ some number $\alpha > 0$ or $\leq$ some number $\beta < 0$ throughout $[-1, 1]$. Consider the operator*

$$L = \sum_{j=h}^{k} a_j(x) \left[\frac{d^j}{dx^j}\right] \tag{7.1}$$

and suppose, throughout $[-1, 1]$,

$$L(f) \geq 0. \tag{7.2}$$

Then, for every integer $n \geq 1$, there is a real polynomial $Q_n(x)$ of degree $\leq n$ such that

$$L(Q_n) \geq 0 \text{ throughout } [-1,1] \tag{7.3}$$

and

$$\max_{-1\leq x\leq 1} |f(x) - Q_n(x)| \leq Cn^{k-p}\omega_1\left(f^{(p)}, \frac{1}{n}\right), \tag{7.4}$$

where C is independent of n or f.

The purpose of this chapter is to extend completely Theorem 7.1 to the fractional level. All involved ordinary derivatives will become now fractional derivatives and even more we will have fractional simultaneous approximation.

We need and make

Definition 7.1. ([4], p. 50) Let $\alpha > 0$ and $\lceil \alpha \rceil = m$, ($\lceil \cdot \rceil$ ceiling of the number). Consider $f \in AC^m([0,1])$ (space of functions f with $f^{(m-1)} \in AC([0,1])$, absolutely continuous functions), $z \in [0,1]$. We define the left Caputo fractional derivative of f of order α as follows:

$$(D_{*z}^{\alpha} f)(x) = \frac{1}{\Gamma(m-\alpha)} \int_z^x (x-t)^{m-\alpha-1} f^{(m)}(t)\,dt, \tag{7.5}$$

for any $x \in [z,1]$, where Γ is the gamma function.

We set

$$D_{*z}^{0} f(x) = f(x),$$

$$D_{*z}^{m} f(x) = f^{(m)}(x), \quad \forall\, x \in [z,1]. \tag{7.6}$$

Definition 7.2. ([5]) Let $\alpha > 0$ and $\lceil \alpha \rceil = m$. Consider $f \in AC^m([0,1])$, $z \in [0,1]$. We define the right Caputo fractional derivative of f of order α as follows:

$$(D_{z-}^{\alpha} f)(x) = \frac{(-1)^m}{\Gamma(m-\alpha)} \int_x^z (t-x)^{m-\alpha-1} f^{(m)}(t)\,dt, \tag{7.7}$$

for any $x \in [0,z]$.

We set

$$D_{z-}^{0} f(x) = f(x),$$

$$D_{z-}^{m} f(x) = (-1)^m f^{(m)}(x), \quad \forall\, x \in [0,z]. \tag{7.8}$$

Remark 7.1. (to Definitions 7.1, 7.2) Let $n \in \mathbb{N}$ with $f^{(n)} \in AC^m([0,1])$, where $\alpha > 0$, $\lceil \alpha \rceil = m$, with $\alpha \notin \mathbb{N}$, here $\lceil n+\alpha \rceil = n + \lceil \alpha \rceil = n+m$, then

$$\left(D_{*z}^{\alpha} f^{(n)}\right)(x) = \frac{1}{\Gamma(m-\alpha)} \int_z^x (x-t)^{m-\alpha-1} \left(f^{(n)}(t)\right)^{(m)} dt$$

$$= \frac{1}{\Gamma((n+m)-(n+\alpha))} \int_z^x (x-t)^{(n+m)-(n+\alpha)-1} f^{(n+m)}(t)\,dt = D_{*z}^{n+\alpha} f(x). \tag{7.9}$$

That is

$$\left(D_{*z}^{\alpha}f^{(n)}\right)(x)=D_{*z}^{n+\alpha}f(x),\quad\forall\, x\in[z,1]. \tag{7.10}$$

Similarly we get

$$\left(D_{z-}^{\alpha}f^{(n)}\right)(x)=\frac{(-1)^{m}}{\Gamma(m-\alpha)}\int_{x}^{z}(t-x)^{m-\alpha-1}\left(f^{(n)}(t)\right)^{(m)}dt$$

$$=\frac{(-1)^{n+m}(-1)^{n}}{\Gamma((n+m)-(n+\alpha))}\int_{x}^{z}(t-x)^{(n+m)-(n+\alpha)-1}f^{(n+m)}(t)dt=(-1)^{n}D_{z-}^{n+\alpha}f(z). \tag{7.11}$$

That is

$$\left(D_{z-}^{\alpha}f^{(n)}\right)(x)=(-1)^{n}D_{z-}^{n+\alpha}f(z),\quad\forall\, x\in[0,z]. \tag{7.12}$$

We need the following:

Consider $f\in C([0,1])$ and the Bernstein polynomials $(B_{N}f)(t)=\sum_{k=0}^{N}f\left(\frac{k}{N}\right)\binom{N}{k}t^{k}(1-t)^{N-k}$, $\forall\, t\in[0,1]$, $N\in\mathbb{N}$, of degree N.

We have $B_{N}1=1$, and B_{N} are positive linear operators.

Theorem 7.2. *([1]) Let $0<\alpha<1$, $r>0$ and $f\in AC([0,1])$ such that $f'\in L_{\infty}([0,1])$.*

Then we have

$$\|B_{N}f-f\|_{\infty}\leq\frac{1}{\Gamma(\alpha+1)}\left(1+\frac{1}{(\alpha+1)r}\right)$$

$$\left[\sup_{x\in[0,1]}\omega_{1}\left(D_{x-}^{\alpha}f,r\left\|B_{n}\left(|\cdot-x|^{\alpha+1}\chi_{[0,x]}(\cdot),x\right)\right\|_{\infty}^{\frac{1}{(\alpha+1)}}\right)_{[0,x]}\right.$$

$$\left\|B_{n}\left(|\cdot-x|^{\alpha+1}\chi_{[0,x]}(\cdot),x\right)\right\|_{\infty}^{\frac{\alpha}{(\alpha+1)}}$$

$$+\sup_{x\in[0,1]}\omega_{1}\left(D_{*x}^{\alpha}f,r\left\|B_{n}\left(|\cdot-x|^{\alpha+1}\chi_{[x,1]}(\cdot),x\right)\right\|_{\infty}^{\frac{1}{(\alpha+1)}}\right)_{[x,1]}$$

$$\left.\left\|B_{n}\left(|\cdot-x|^{\alpha+1}\chi_{[x,1]}(\cdot),x\right)\right\|_{\infty}^{\frac{\alpha}{(\alpha+1)}}\right],\quad\forall\, N\in\mathbb{N}. \tag{7.13}$$

Above χ stands for the characteristic function, also the two first moduli of continuity are over the intervals $[0,x]$ and $[x,1]$, respectively as indicated.

Remark 7.2. (to Theorem 7.2) Next we choose $r = \frac{1}{\alpha+1}$, $p = \frac{2}{\alpha+1} > 1$, $q = \frac{2}{1-\alpha} > 1$ with $\frac{1}{p} + \frac{1}{q} = 1$.

We observe that both

$$B_n\left(|\cdot - x|^{\alpha+1}\chi_{[0,x]}(\cdot), x\right), \; B_n\left(|\cdot - x|^{\alpha+1}\chi_{[x,1]}(\cdot), x\right)$$

$$\leq B_n\left(|\cdot - x|^{\alpha+1}, x\right) = \sum_{k=0}^{N}\left|x - \frac{k}{N}\right|^{\alpha+1}\binom{N}{k}x^k(1-x)^{N-k} \tag{7.14}$$

(by discrete Hölder's inequality)

$$\leq \left(\sum_{k=0}^{N}\left(x - \frac{k}{N}\right)^2\binom{N}{k}x^k(1-x)^{N-k}\right)^{\frac{\alpha+1}{2}} = \left(\frac{x(1-x)}{N}\right)^{\frac{\alpha+1}{2}} \tag{7.15}$$

$$\leq \frac{1}{(4N)^{\frac{\alpha+1}{2}}} = \frac{1}{\left(2\sqrt{N}\right)^{\alpha+1}}, \quad \forall\, x \in [0,1].$$

We have proved the following important auxiliary result.

Theorem 7.3. *Let $0 < \alpha < 1$, $f \in AC([0,1])$, with $f' \in L_\infty([0,1])$, $N \in \mathbb{N}$. Then*

$$\|B_N f - f\|_\infty \leq \frac{2^{1-\alpha}}{\Gamma(\alpha+1)N^{\frac{\alpha}{2}}}\left[\sup_{x\in[0,1]}\omega_1\left(D_{x-}^{\alpha}f, \frac{1}{2(\alpha+1)N^{\frac{1}{2}}}\right)_{[0,x]} + \sup_{x\in[0,1]}\omega_1\left(D_{*x}^{\alpha}f, \frac{1}{2(\alpha+1)N^{\frac{1}{2}}}\right)_{[x,1]}\right] =: T_N^{\alpha}(f) < \infty. \tag{7.16}$$

Proof. By (7.13) and (7.15).

By [1] we get that the quantity within the bracket of (7.16) is finite. □

So as $N \to \infty$ we derive $B_N f \xrightarrow{u} f$ (uniformly) with rates.

7.2 Main Result

We give the following simultaneous approximation fractional result.

Theorem 7.4. *Let $\beta > 0$, $\beta \notin \mathbb{N}$, with integral part $[\beta] = n \in \mathbb{Z}_+$ such that $\beta = n + \alpha$, where $0 < \alpha < 1$. Let $f \in AC^{n+1}([0,1])$, and $f^{(n+1)} \in L_\infty([0,1])$, $N \in \mathbb{N}$. Set*

$$P_{N+n}(f)(x) := \sum_{k=0}^{n-1}\frac{f^{(k)}(0)}{k!}x^k + \int_0^x\left(\int_0^{x_{n-1}}\ldots\left(\int_0^{x_1}B_N\left(f^{(n)}\right)(t_1)\,dt_1\right)dx_1\ldots\right)dx_{n-1} \tag{7.17}$$

$$= \sum_{k=0}^{n-1} \frac{f^{(k)}(0)}{k!} x^k + \frac{1}{(n-1)!} \int_0^x (x-t)^{n-1} B_N\left(f^{(n)}\right)(t)\,dt,$$

for all $0 \leq x \leq 1$*, a polynomial of degree* $(N+n)$*. Here* $B_N\left(f^{(n)}\right)$ *is the Bernstein polynomial of degree* N*. If n=0 the sum in (7.17) collapses.*

Set also

$$T_N^{\beta,\alpha}(f) := \frac{2^{1-\alpha}}{\Gamma(\alpha+1) N^{\frac{\alpha}{2}}} \left[\sup_{x\in[0,1]} \omega_1 \left(D_{x-}^{\beta} f, \frac{1}{2(\alpha+1) N^{\frac{1}{2}}} \right)_{[0,x]} \right.$$

$$\left. + \sup_{x\in[0,1]} \omega_1 \left(D_{*x}^{\beta} f, \frac{1}{2(\alpha+1) N^{\frac{1}{2}}} \right)_{[x,1]} \right] < \infty, \tag{7.18}$$

for every $N \in \mathbb{N}$*.*

Then $P_{N+n}^{(i)}(f) = P_{N+n-i}(f)$*, and*

$$\left\| P_{N+n}^{(i)}(f) - f^{(i)} \right\|_{\infty,[0,1]} \leq \frac{T_N^{\beta,\alpha}(f)}{(n-i)!}, \quad i = 0, 1, ..., n. \tag{7.19}$$

As $N \to \infty$ *we derive with rates* $P_{N+n}^{(i)}(f) \xrightarrow{u} f^{(i)}$*.*

Proof. Notice here $\lceil \beta \rceil = (n+1) \in \mathbb{N}$. We have by (7.16) that

$$\left\| B_N\left(f^{(n)}\right) - f^{(n)} \right\|_{\infty,[0,1]} \leq T_N^{\alpha}\left(f^{(n)}\right) < \infty. \tag{7.20}$$

that is

$$-T_N^{\alpha}\left(f^{(n)}\right) \leq B_N\left(f^{(n)}\right)(t_1) - f^{(n)}(t_1) \leq T_N^{\alpha}\left(f^{(n)}\right), \tag{7.21}$$

for every $0 \leq t_1 \leq x_1 \leq 1$.

Set

$$P_N(f)(x) := B_N\left(f^{(n)}\right)(x). \tag{7.22}$$

Hence it holds

$$-T_N^{\alpha}\left(f^{(n)}\right) x_1 \leq \int_0^{x_1} B_N\left(f^{(n)}\right)(t_1)\,dt_1 - \int_0^{x_1} f^{(n)}(t_1)\,dt_1 \leq T_N^{\alpha}\left(f^{(n)}\right) x_1, \tag{7.23}$$

that is

$$-T_N^{\alpha}\left(f^{(n)}\right) x_1 \leq f^{(n-1)}(0) + \int_0^{x_1} B_N\left(f^{(n)}\right)(t_1)\,dt_1$$

$$-f^{(n-1)}(x_1) \leq T_N^{\alpha}\left(f^{(n)}\right) x_1. \tag{7.24}$$

Set

$$P_{N+1}(f)(x) := f^{(n-1)}(0) + \int_0^x B_N\left(f^{(n)}\right)(t_1)\,dt_1, \tag{7.25}$$

that is

$$P'_{N+1}(f)(x) = P_N(f)(x), \quad \text{all } x \in [0,1]. \tag{7.26}$$

Hence

$$-T_N^{\alpha}\left(f^{(n)}\right)x_1 \leq P_{N+1}(f)(x_1) - f^{(n-1)}(x_1) \leq T_N^{\alpha}\left(f^{(n)}\right)x_1, \tag{7.27}$$

for all $x_1 \in [0,1]$.

Continuing like this we get

$$-T_N^{\alpha}\left(f^{(n)}\right)\frac{x_2^2}{2} \leq f^{(n-2)}(0) + \int_0^{x_2} P_{N+1}(f)(x_1)\,dx_1$$

$$-f^{(n-2)}(x_2) \leq T_N^{\alpha}\left(f^{(n)}\right)\frac{x_2^2}{2}, \tag{7.28}$$

all $0 \leq x_1 \leq x_2 \leq 1$.

Set

$$P_{N+2}(f)(x) := f^{(n-2)}(0) + \int_0^{x} P_{N+1}(f)(x_1)\,dx_1, \tag{7.29}$$

that is

$$P'_{N+2}(f)(x) = P_{N+1}(f)(x), \tag{7.30}$$

and

$$P''_{N+2}(f)(x) = P_N(f)(x), \quad \text{all } x \in [0,1]. \tag{7.31}$$

So far we have

$$-T_N^{\alpha}\left(f^{(n)}\right)\frac{x_2^2}{2} \leq P_{N+2}(f)(x_2) - f^{(n-2)}(x_2) \leq T_N^{\alpha}\left(f^{(n)}\right)\frac{x_2^2}{2}, \tag{7.32}$$

for all $x_2 \in [0,1]$.

Similarly we derive

$$-T_N^{\alpha}\left(f^{(n)}\right)\frac{x_3^3}{3!} \leq f^{(n-3)}(0) + \int_0^{x_3} P_{N+2}(f)(x_2)\,dx_2$$

$$-f^{(n-3)}(x_3) \leq T_N^{\alpha}\left(f^{(n)}\right)\frac{x_3^3}{3!}, \tag{7.33}$$

all $0 \leq x_2 \leq x_3 \leq 1$.

Set

$$P_{N+3}(f)(x) := f^{(n-3)}(0) + \int_0^{x} P_{N+2}(f)(x_2)\,dx_2, \tag{7.34}$$

that is

$$P'_{N+3}(f)(x) = P_{N+2}(f)(x), \tag{7.35}$$

and

$$P'''_{N+3}(f)(x) = P_N(f)(x), \quad \text{all } x \in [0,1]. \tag{7.36}$$

Hence

$$-T_N^{\alpha}\left(f^{(n)}\right)\frac{x_3^3}{3!}\leq P_{N+3}(f)(x_3)-f^{(n-3)}(x_3)\leq T_N^{\alpha}\left(f^{(n)}\right)\frac{x_3^3}{3!},\tag{7.37}$$

for all $x_3\in[0,1]$.

Continuing as above, after n steps, we derive

$$-T_N^{\alpha}\left(f^{(n)}\right)\frac{x_n^n}{n!}\leq P_{N+n}(f)(x_n)-f(x_n)\leq T_N^{\alpha}\left(f^{(n)}\right)\frac{x_n^n}{n!},\tag{7.38}$$

with $0\leq x_n\leq 1$.

Above

$$P_{N+n}(f)(x):=f(0)+\int_0^x P_{N+n-1}(f)(x_{n-1})\,dx_{n-1},\tag{7.39}$$

that is

$$P'_{N+n}(f)(x)=P_{N+n-1}(f)(x),\tag{7.40}$$

and

$$P_{N+n}^{(n)}(f)(x)=P_N(f)(x)=B_N\left(f^{(n)}\right)(x),\quad \text{all } x\in[0,1].\tag{7.41}$$

So clearly here $P_{N+n}(f)$ has the representations (7.17), the second one comes by Taylor's theorem.

By (7.21) we get

$$\left\|P_N(f)-f^{(n)}\right\|_\infty\leq T_N^{\alpha}\left(f^{(n)}\right),\tag{7.42}$$

by (7.27) we find

$$\left\|P_{N+1}(f)-f^{(n-1)}\right\|_\infty\leq T_N^{\alpha}\left(f^{(n)}\right),\tag{7.43}$$

by (7.32) we derive

$$\left\|P_{N+2}(f)-f^{(n-2)}\right\|_\infty\leq \frac{T_N^{\alpha}\left(f^{(n)}\right)}{2},\tag{7.44}$$

by (7.37) we obtain

$$\left\|P_{N+3}(f)-f^{(n-3)}\right\|_\infty\leq \frac{T_N^{\alpha}\left(f^{(n)}\right)}{3!},\tag{7.45}$$

and by (7.38) we have

$$\|P_{N+n}(f)-f\|_\infty\leq \frac{T_N^{\alpha}\left(f^{(n)}\right)}{n!}.\tag{7.46}$$

So we have proved that

$$\left\|P_{N+n}^{(i)}(f)-f^{(i)}\right\|_\infty\leq \frac{T_N^{\alpha}\left(f^{(n)}\right)}{(n-i)!},\quad i=0,1,...,n.\tag{7.47}$$

Based on (7.10) and (7.12) we derive that

$$T_n^{\alpha}\left(f^{(n)}\right)=T_N^{\beta,\alpha}(f).\tag{7.48}$$

The quantity within the bracket of (7.18), by [1], is finite.

The proof of the theorem now is complete. □

We completely left fractionalize Theorem 7.4, to have

Theorem 7.5. *Here all terms and assumptions are as in Theorem 7.4. Consider* $\alpha_j > 0$, $j = 1, ..., n \in \mathbb{N}$, *such that* $\alpha_0 = 0 < \alpha_1 \leq 1 < \alpha_2 \leq 2 < \alpha_3 \leq 3 < ... < ... < \alpha_n \leq n$.

Then

$$\left\| D_{*0}^{\alpha_j}(f) - D_{*0}^{\alpha_j}(P_{N+n}(f)) \right\|_{\infty,[0,1]} \leq \frac{T_N^{\beta,\alpha}(f)}{\Gamma(j-\alpha_j+1)(n-j)!}, \quad j = 0, 1, ..., n. \tag{7.49}$$

Notice (7.49) generalizes (7.19).

Proof. Let $\alpha_j > 0$, $j = 1, ..., n$, such that $0 < \alpha_1 \leq 1 < \alpha_2 \leq 2 < \alpha_3 \leq 3 < ... < ... < \alpha_n \leq n$. That is $\lceil \alpha_j \rceil = j$, $j = 1, ..., n$.

We consider the left Caputo fractional derivatives

$$\left(D_*^{\alpha_j} f\right)(x) = \frac{1}{\Gamma(j-\alpha_j)} \int_0^x (x-t)^{j-\alpha_j-1} f^{(j)}(t)\, dt, \tag{7.50}$$

$$\left(D_{*0}^{j} f\right)(x) = f^{(j)}(x),$$

and

$$\left(D_{*0}^{\alpha_j}(P_{N+n}(f))\right)(x) = \frac{1}{\Gamma(j-\alpha_j)} \int_0^x (x-t)^{j-\alpha_j-1} (P_{N+n}(f))^{(j)}(t)\, dt,$$

$$\left(D_{*0}^{j}(P_{N+n}(f))\right)(x) = (P_{N+n}(f))^{(j)}. \tag{7.51}$$

We notice that

$$\left| \left(D_{*0}^{\alpha_j} f\right)(x) - \left(D_{*0}^{\alpha_j}(P_{N+n}(f))\right)(x) \right|$$

$$= \frac{1}{\Gamma(j-\alpha_j)} \left| \int_0^x (x-t)^{j-\alpha_j-1} f^{(j)}(t)\, dt - \int_0^x (x-t)^{j-\alpha_j-1} (P_{N+n}(f))^{(j)}(t)\, dt \right|$$

$$= \frac{1}{\Gamma(j-\alpha_j)} \left| \int_0^x (x-t)^{j-\alpha_j-1} \left(f^{(j)}(t) - (P_{N+n}(f))^{(j)}(t) \right) dt \right|$$

$$\leq \frac{1}{\Gamma(j-\alpha_j)} \int_0^x (x-t)^{j-\alpha_j-1} \left| f^{(j)}(t) - (P_{N+n}(f))^{(j)}(t) \right| dt \tag{7.52}$$

$$\overset{(7.19)}{\leq} \frac{1}{\Gamma(j-\alpha_j)} \left(\int_0^x (x-t)^{j-\alpha_j-1} dt \right) \frac{T_N^{\beta,\alpha}(f)}{(n-j)!}$$

$$= \frac{x^{j-\alpha_j}}{\Gamma(j-\alpha_j)(j-\alpha_j)} \frac{T_N^{\beta,\alpha}(f)}{(n-j)!} = \frac{x^{j-\alpha_j}}{\Gamma(j-\alpha_j+1)} \frac{T_N^{\beta,\alpha}(f)}{(n-j)!}. \tag{7.53}$$

We have proved

$$\left| \left(D_{*0}^{\alpha_j} f\right)(x) - \left(D_{*0}^{\alpha_j}(P_{N+n}(f))\right)(x) \right| \leq \frac{x^{j-\alpha_j} T_N^{\beta,\alpha}(f)}{\Gamma(j-\alpha_j+1)(n-j)!}$$

$$\leq \frac{T_N^{\beta,\alpha}(f)}{\Gamma(j-\alpha_j+1)(n-j)!}, \tag{7.54}$$

for every $x \in [0,1]$, proving the claim. □

We completely right fractionalize Theorem 7.4, to have

Theorem 7.6. *Here all terms and assumptions are as in Theorem 7.5. It holds*

$$\left\| D_{1-}^{\alpha_j}(f) - D_{1-}^{\alpha_j}(P_{N+n}(f)) \right\|_{\infty,[0,1]} \leq \frac{T_N^{\beta,\alpha}(f)}{\Gamma(j-\alpha_j+1)(n-j)!}, \tag{7.55}$$

$j = 0, 1, ..., n.$

Observe that (7.55) generalizes (7.19).

Proof. We notice that

$$\left| \left(D_{1-}^{\alpha_j} f \right)(x) - \left(D_{1-}^{\alpha_j} (P_{N+n}(f)) \right)(x) \right|$$

$$= \frac{1}{\Gamma(j-\alpha_j)} \left| \int_x^1 (t-x)^{j-\alpha_j-1} f^{(j)}(t)\, dt - \int_x^1 (t-x)^{j-\alpha_j-1} (P_{N+n}(f))^{(j)}(t)\, dt \right|$$

$$= \frac{1}{\Gamma(j-\alpha_j)} \left| \int_x^1 (t-x)^{j-\alpha_j-1} \left(f^{(j)}(t) - (P_{N+n}(f))^{(j)}(t) \right) dt \right| \tag{7.56}$$

$$\leq \frac{1}{\Gamma(j-\alpha_j)} \int_x^1 (t-x)^{j-\alpha_j-1} \left| f^{(j)}(t) - (P_{N+n}(f))^{(j)}(t) \right| dt$$

$$\overset{(7.19)}{\leq} \frac{1}{\Gamma(j-\alpha_j)} \left(\int_x^1 (t-x)^{j-\alpha_j-1} dt \right) \frac{T_N^{\beta,\alpha}(f)}{(n-j)!}$$

$$= \frac{(1-x)^{j-\alpha_j}}{\Gamma(j-\alpha_j+1)} \frac{T_N^{\beta,\alpha}(f)}{(n-j)!}. \tag{7.57}$$

We have proved

$$\left| \left(D_{1-}^{\alpha_j} f \right)(x) - \left(D_{1-}^{\alpha_j} (P_{N+n}(f)) \right)(x) \right| \leq \frac{(1-x)^{j-\alpha_j}}{\Gamma(j-\alpha_j+1)} \frac{T_N^{\beta,\alpha}(f)}{(n-j)!}$$

$$\leq \frac{T_N^{\beta,\alpha}(f)}{\Gamma(j-\alpha_j+1)(n-j)!}, \tag{7.58}$$

for all $x \in [0,1]$, proving the claim. □

It follows to important

Corollary 7.1. *Here all terms and assumptions are as in Theorem 7.5. Let $\lambda \in [0,1]$. Then*

$$\Big\| \left(\lambda D_{*0}^{\alpha_j}(f) + (1-\lambda) D_{1-}^{\alpha_j}(f) \right)$$

$$- \left(\lambda D_{*0}^{\alpha_j}(P_{N+n}(f)) + (1-\lambda) D_{1-}^{\alpha_j}(P_{N+n}(f)) \right) \Big\|_{\infty,[0,1]}$$

$$\leq \frac{T_N^{\beta,\alpha}(f)}{\Gamma(j-\alpha_j+1)(n-j)!}, \quad j = 0, 1, ..., n. \tag{7.59}$$

Proof. We see that

$$\left\|\left(\lambda D_{*0}^{\alpha_j}(f)+(1-\lambda) D_{1-}^{\alpha_j}(f)\right)\right.$$
$$\left.-\left(\lambda D_{*0}^{\alpha_j}\left(P_{N+n}(f)\right)+(1-\lambda) D_{1-}^{\alpha_j}\left(P_{N+n}(f)\right)\right)\right\|_{\infty,[0,1]}$$
$$=\left\|\lambda\left(D_{*0}^{\alpha_j}(f)-D_{*0}^{\alpha_j}\left(P_{N+n}(f)\right)\right)+(1-\lambda)\left(D_{1-}^{\alpha_j}(f)-D_{1-}^{\alpha_j}\left(P_{N+n}(f)\right)\right)\right\|_{\infty} \tag{7.60}$$

$$\leq \lambda\left\|D_{*0}^{\alpha_j}(f)-D_{*0}^{\alpha_j}\left(P_{N+n}(f)\right)\right\|_{\infty}$$
$$+(1-\lambda)\left\|D_{1-}^{\alpha_j}(f)-D_{1-}^{\alpha_j}\left(P_{N+n}(f)\right)\right\|_{\infty}$$

$$\stackrel{((7.49),(7.55))}{\leq} \lambda \frac{T_N^{\beta,\alpha}(f)}{\Gamma\left(j-\alpha_j+1\right)(n-j)!}+(1-\lambda) \frac{T_N^{\beta,\alpha}(f)}{\Gamma\left(j-\alpha_j+1\right)(n-j)!} \tag{7.61}$$

$$=\frac{T_N^{\beta,\alpha}(f)}{\Gamma\left(j-\alpha_j+1\right)(n-j)!},$$

proving the claim. □

Next comes our main result: the complete fractional simultaneous monotone uniform approximation, using mixed fractional differential operators.

Theorem 7.7. *Let $\beta>0$, $\beta \notin \mathbb{N}$, $n=[\beta] \in \mathbb{N}: \beta=n+\alpha$, $0<\alpha<1$; $f \in AC^{n+1}([0,1])$, and $f^{(n+1)} \in L_{\infty}([0,1])$. Let $\lambda \in[0,1]$, and $h, k \in \mathbb{Z}_{+}$ with $0 \leq h \leq k \leq n$, when $\lambda \neq 1$ we take h to be even. Consider the numbers $\alpha_0=0<\alpha_1 \leq 1<\alpha_2 \leq 2<\alpha_3 \leq 3<\ldots<\ldots<\alpha_n \leq n$. Let $\alpha_j(x)$, $j=h, h+1, \ldots, k$ be real functions, defined and bounded on $[0,1]$, and suppose $\alpha_h(x)$ is either $\geq \overline{\alpha}>0$ or $\leq \overline{\beta}<0$ on $[0,1]$. We set*

$$l_\tau :\equiv \sup_{x \in[0,1]}\left|\alpha_h^{-1}(x) \alpha_\tau(x)\right|, \quad \tau=h, \ldots, k. \tag{7.62}$$

Here $T_N^{\beta,\alpha}(f)$, $N \in \mathbb{N}$, as in (7.18), $(N \rightarrow \infty)$. Consider the mixed fractional linear differential operator

$$L^*:=\sum_{j=h}^k \alpha_j(x)\left[\lambda D_{*0}^{\alpha_j}+(1-\lambda) D_{1-}^{\alpha_j}\right]. \tag{7.63}$$

Then, for any $N \in \mathbb{N}$, there exists a real polynomial Q_{N+n} of degree $(N+n)$ such that

1) for $j=h+1, \ldots, n$, it holds

$$\left\|\left(\lambda D_{*0}^{\alpha_j}(f(x))+(1-\lambda) D_{1-}^{\alpha_j}(f(x))\right)\right.$$

$$\left.-\left(\lambda D_{*0}^{\alpha_j}\left(Q_{N+n}(x)\right)+(1-\lambda) D_{1-}^{\alpha_j}\left(Q_{N+n}(x)\right)\right)\right\|_{\infty,[0,1]}$$

$$\leq \frac{T_N^{\beta,\alpha}(f)}{\Gamma(j-\alpha_j+1)(n-j)!}, \tag{7.64}$$

2) for $j = 1, ..., h$, it holds

$$\left\| \left(\lambda D_{*0}^{\alpha_j}(f) + (1-\lambda) D_{1-}^{\alpha_j}(f)\right) - \left(\lambda D_{*0}^{\alpha_j}(Q_{N+n}) + (1-\lambda) D_{1-}^{\alpha_j}(Q_{N+n})\right) \right\|_{\infty,[0,1]} \tag{7.65}$$

$$\leq \frac{T_N^{\beta,\alpha}(f)}{(h-j)!} \left[\frac{1}{\Gamma(j-\alpha_j+1)} + \left(\sum_{\tau=h}^{k} \frac{l_\tau}{\Gamma(\tau-\alpha_\tau+1)(n-\tau)!} \right) \right.$$

$$\left. \left\{ \lambda \frac{\Gamma(h-j+1)}{\Gamma(h-\alpha_j+1)} + (1-\lambda) \left[\sum_{\theta=0}^{h-j} \binom{h-j}{\theta} \frac{\Gamma(h-j-\theta+1)}{\Gamma(h-\alpha_j-\theta+1)} \right] \right\} \right],$$

and

3)

$$\|f - Q_{N+n}\|_{\infty,[0,1]} \leq \frac{T_N^{\beta,\alpha}(f)}{h!} \left(\sum_{\tau=h}^{k} \frac{l_\tau}{\Gamma(\tau-\alpha_\tau+1)(n-\tau)!} + 1 \right). \tag{7.66}$$

The set

$$\Lambda := \left\{ (\lambda, x) \in (0,1)^2 : \lambda x^{h-\alpha_h} + (1-\lambda)(1-x)^{h-\alpha_h} \geq \Gamma(h-\alpha_h+1) \right\} \tag{7.67}$$

is not empty.

1) We assume that $L^* f(x) \geq 0$, for every $x \in \Lambda$, then $L^*(Q_{N+n}) \geq 0$.

2) If $L^* f(0) \geq 0$ and $0 \leq \lambda \leq 1 - \Gamma(h-\alpha_h+1)$, then $L^*(Q_{N+n})(0) \geq 0$.

3) If $L^* f(1) \geq 0$ and $\Gamma(h-\alpha_h+1) \leq \lambda \leq 1$, then $L^*(Q_{N+n})(1) \geq 0$.

4) Given $L^* f(0) \geq 0$, $\lambda = 0$, we get $L^*(Q_{N+n})(0) \geq 0$.

5) Given $L^* f(1) \geq 0$, $\lambda = 1$, we derive $L^*(Q_{N+n})(1) \geq 0$.

6) Let $\lambda = 0$, h even, $h > \alpha_h$, $L^* f(x) \geq 0$, for $x \in (0,1)$ such that $x \leq 1 - \Gamma(h-\alpha_h+1)^{\frac{1}{h-\alpha_h}}$, then $L^*(Q_{N+n})(x) \geq 0$.

Finally:

7) Let $\lambda = 1$, $h > \alpha_h$, and $x \in (0,1)$ with $x \geq (\Gamma(h-\alpha_h+1))^{\frac{1}{h-\alpha_h}}$. Assume there $L^* f(x) \geq 0$, then $L^*(Q_{N+n})(x) \geq 0$.

Proof. Here let h, k be integers $0 \leq h \leq k \leq n$, and $\alpha_0 = 0 < \alpha_1 \leq 1 < \alpha_2 \leq 2 < \alpha_3 \leq 3 < ... < ... < \alpha_n \leq n$, that is $\lceil \alpha_j \rceil = j$, $j = 1, ..., n$. We set

$$l_{j_*} :\equiv \sup_{x \in [0,1]} \left| \alpha_h^{-1}(x) \alpha_{j_*}(x) \right| < \infty, \quad h \leq j_* \leq k, \tag{7.68}$$

and

$$\rho_N := T_N^{\beta,\alpha}(f) \left(\sum_{j_*=h}^{k} \frac{l_{j_*}}{\Gamma(j_*-\alpha_{j_*}+1)(n-j_*)!} \right). \tag{7.69}$$

I. Suppose, throughout $[0,1]$, $\alpha_h(x) \geq \overline{\alpha} > 0$. Call

$$Q_{N+n}(x) := P_{N+n}(f)(x) + \rho_N \frac{x^h}{h!}, \tag{7.70}$$

where $P_{N+n}(f)(x)$ as in (7.17).

Then by (7.59) we obtain

$$\left\| \left(\lambda D_{*0}^{\alpha_j} + (1-\lambda) D_{1-}^{\alpha_j}\right) \left(f(x) + \rho_N \frac{x^h}{h!}\right) - \left(\lambda D_{*0}^{\alpha_j} + (1-\lambda) D_{1-}^{\alpha_j}\right) (Q_{N+n}(x)) \right\|_{\infty,[0,1]}$$

$$\leq \frac{T_N^{\beta,\alpha}(f)}{\Gamma(j-\alpha_j+1)(n-j)!}, \tag{7.71}$$

all $0 \leq j \leq n$.

When $h+1 \leq j \leq n$, immediately by (7.71) we obtain

$$\left\| \left(\lambda D_{*0}^{\alpha_j} + (1-\lambda) D_{1-}^{\alpha_j}\right) (f(x)) - \left(\lambda D_{*0}^{\alpha_j} + (1-\lambda) D_{1-}^{\alpha_j}\right) (Q_{N+n}(x)) \right\|_{\infty,[0,1]}$$

$$\leq \frac{T_N^{\beta,\alpha}(f)}{\Gamma(j-\alpha_j+1)(n-j)!}, \tag{7.72}$$

proving (7.64).

Next we treat the case of $1 \leq j \leq h$. We have after calculations that

$$D_{*0}^{\alpha_j}\left(\frac{x^h}{h!}\right) = \frac{\Gamma(h-j+1)\, x^{h-\alpha_j}}{\Gamma(h-\alpha_j+1)(h-j)!}, \tag{7.73}$$

and

$$D_{1-}^{\alpha_j}\left(\frac{x^h}{h!}\right) = \frac{1}{(h-j)!} \cdot \left[\sum_{\theta=0}^{h-j} \binom{h-j}{\theta} (-1)^{h+\theta} \left\{\frac{\Gamma(h-j-\theta+1)}{\Gamma(h-\alpha_j-\theta+1)} (1-x)^{h-\alpha_j-\theta}\right\}\right]. \tag{7.74}$$

Hence by (7.71) we have

$$\Bigg\| \left(\lambda D_{*0}^{\alpha_j} + (1-\lambda) D_{1-}^{\alpha_j}\right) (f(x)) + \rho_N \left\{\lambda \frac{\Gamma(h-j+1)\, x^{h-\alpha_j}}{\Gamma(h-\alpha_j+1)(h-j)!} + \frac{(1-\lambda)}{(h-j)!} \left[\sum_{\theta=0}^{h-j} \binom{h-j}{\theta} (-1)^{h+\theta} \left\{\frac{\Gamma(h-j-\theta+1)}{\Gamma(h-\alpha_j-\theta+1)} (1-x)^{h-\alpha_j-\theta}\right\}\right]\right\} - \left(\lambda D_{*0}^{\alpha_j} + (1-\lambda) D_{1-}^{\alpha_j}\right) (Q_{N+n}(x)) \Bigg\|_{\infty,[0,1]} \leq \frac{T_N^{\beta,\alpha}(f)}{\Gamma(j-\alpha_j+1)(n-j)!}, \tag{7.75}$$

all $1 \leq j \leq h$.

By (7.75) and triangle inequality we get

$$\left\| \left(\lambda D_{*0}^{\alpha_j} + (1-\lambda) D_{1-}^{\alpha_j}\right) (f) - \left(\lambda D_{*0}^{\alpha_j} + (1-\lambda) D_{1-}^{\alpha_j}\right) (Q_{N+n}) \right\|_{\infty,[0,1]}$$

$$\leq \frac{T_N^{\beta,\alpha}(f)}{\Gamma(j-\alpha_j+1)(n-j)!}$$
$$+\frac{\rho_N}{(h-j)!}\left\{\lambda\frac{\Gamma(h-j+1)}{\Gamma(h-\alpha_j+1)}+(1-\lambda)\left[\sum_{\theta=0}^{h-j}\binom{h-j}{\theta}\frac{\Gamma(h-j-\theta+1)}{\Gamma(h-\alpha_j-\theta+1)}\right]\right\} \tag{7.76}$$

$$= \frac{T_N^{\beta,\alpha}(f)}{\Gamma(j-\alpha_j+1)(n-j)!}+\frac{T_N^{\beta,\alpha}(f)}{(h-j)!}\left(\sum_{j_*=h}^{k}\frac{l_{j_*}}{\Gamma(j_*-\alpha_{j_*}+1)(n-j_*)!}\right)$$
$$\cdot\left\{\lambda\frac{\Gamma(h-j+1)}{\Gamma(h-\alpha_j+1)}+(1-\lambda)\left[\sum_{\theta=0}^{h-j}\binom{h-j}{\theta}\frac{\Gamma(h-j-\theta+1)}{\Gamma(h-\alpha_j-\theta+1)}\right]\right\}$$
$$\leq \frac{T_N^{\beta,\alpha}(f)}{(h-j)!}\left[\frac{1}{\Gamma(j-\alpha_j+1)}+\left(\sum_{\tau=h}^{k}\frac{l_\tau}{\Gamma(\tau-\alpha_\tau+1)(n-\tau)!}\right)\right.$$
$$\left.\cdot\left\{\lambda\frac{\Gamma(h-j+1)}{\Gamma(h-\alpha_j+1)}+(1-\lambda)\left[\sum_{\theta=0}^{h-j}\binom{h-j}{\theta}\frac{\Gamma(h-j-\theta+1)}{\Gamma(h-\alpha_j-\theta+1)}\right]\right\}\right]=:K. \tag{7.77}$$

So we have derived

$$\left\|\left(\lambda D_{*0}^{\alpha_j}+(1-\lambda)D_{1-}^{\alpha_j}\right)(f)-\left(\lambda D_{*0}^{\alpha_j}+(1-\lambda)D_{1-}^{\alpha_j}\right)(Q_{N+n})\right\|_{\infty,[0,1]}\leq K, \tag{7.78}$$

$j=1,...,h$, proving (7.65).

When $j=0$ from (7.71) we obtain

$$\left\|f(x)+\rho_N\frac{x^h}{h!}-Q_{N+n}(x)\right\|_{\infty,[0,1]}\leq\frac{T_N^{\beta,\alpha}(f)}{n!}. \tag{7.79}$$

Hence

$$\|f-Q_{N+n}\|_{\infty,[0,1]}\leq\frac{\rho_N}{h!}+\frac{T_N^{\beta,\alpha}(f)}{n!}$$
$$=\frac{T_N^{\beta,\alpha}(f)}{h!}\left(\sum_{\tau=h}^{k}\frac{l_\tau}{\Gamma(\tau-\alpha_\tau+1)(n-\tau)!}\right)+\frac{T_N^{\beta,\alpha}(f)}{n!} \tag{7.80}$$
$$\leq\frac{T_N^{\beta,\alpha}(f)}{h!}\left(\sum_{\tau=h}^{k}\frac{l_\tau}{\Gamma(\tau-\alpha_\tau+1)(n-\tau)!}+1\right), \tag{7.81}$$

proving (7.66).

Furthermore, if (λ,x) is in the critical set Λ, see (7.67), and $L^*f(x)\geq 0$, we get

$$\alpha_h^{-1}L^*(Q_{N+n})=\alpha_h^{-1}(x)L^*(f(x))$$

$$+\rho_N \left\{ \lambda \frac{x^{h-\alpha_h}}{\Gamma(h-\alpha_h+1)} + (1-\lambda) \frac{(1-x)^{h-\alpha_h}}{\Gamma(h-\alpha_h+1)} \right\} \tag{7.82}$$

(when $\lambda \in [0,1)$ we assumed that h is even)

$$+\sum_{j=h}^{k} \alpha_h^{-1}(x)\,\alpha_j(x) \left\{ \left(\lambda D_{*0}^{\alpha_j} + (1-\lambda) D_{1-}^{\alpha_j}\right) \left[Q_{N+n}(x) - f(x) - \rho_N \frac{x^h}{h!} \right] \right\}$$

$$\overset{(7.71)}{\geq} \rho_N \left\{ \lambda \frac{x^{h-\alpha_h}}{\Gamma(h-\alpha_h+1)} + (1-\lambda) \frac{(1-x)^{h-\alpha_h}}{\Gamma(h-\alpha_h+1)} \right\} - \sum_{j=h}^{k} l_j \frac{T_N^{\beta,\alpha}(f)}{\Gamma(j-\alpha_j+1)(n-j)!}$$

$$= \rho_N \left\{ \lambda \frac{x^{h-\alpha_h}}{\Gamma(h-\alpha_h+1)} + (1-\lambda) \frac{(1-x)^{h-\alpha_h}}{\Gamma(h-\alpha_h+1)} - 1 \right\}$$

$$= \frac{\rho_N}{\Gamma(h-\alpha_h+1)} \left\{ \lambda x^{h-\alpha_h} + (1-\lambda)(1-x)^{h-\alpha_h} - \Gamma(h-\alpha_h+1) \right\} =: A(x,\lambda). \tag{7.83}$$

The set

$$\Lambda := \left\{ (\lambda, x) \in (0,1)^2 : \lambda x^{h-\alpha_h} + (1-\lambda)(1-x)^{h-\alpha_h} \geq \Gamma(h-\alpha_h+1) \right\} \tag{7.84}$$

is not empty.

If $h = \alpha_h$ then $\Lambda = (0,1)^2$. Assume $\alpha_h < h$:

Let us choose $\lambda = x = \delta \in (0,1)$, some want to find specific examples of

$$\delta^{1+h-\alpha_h} + (1-\delta)^{1+h-\alpha_h} \geq \Gamma(h-\alpha_h+1). \tag{7.85}$$

The minimum value of Γ over $(0,\infty)$ is $\Gamma(1.46163) \simeq 0.885603$, we pick here $1 + h - \alpha_h = 1.46163$ and $\delta = 0.99$.

Hence

$$(0.99)^{1.46163} + (0.01)^{1.46163} \tag{7.86}$$

$$= 0.985417497 + 0.001193274 = 0.986610771 > 0.885603.$$

Similarly, we have that $\Gamma(1.4) = 0.887264$, and we pick $\delta = 0.95$ and $1+h-\alpha_h = 1.4$. Then

$$(0.95)^{1.4} + (0.05)^{1.4}$$

$$= 0.930707144 + 0.015085441 = 0.945792585 > 0.887264. \tag{7.87}$$

Hence $\Lambda \neq \varnothing$.

Hence over Λ we get that $A(x,\lambda) \geq 0$, thus there $L^*(Q_{N+n}) \geq 0$.

We know that $\Gamma(1) = 1$, $\Gamma(2) = 1$, and Γ is convex and positive on $(0,\infty)$. Here in general $0 \leq h - \alpha_h < 1$, hence $1 \leq h - \alpha_h + 1 < 2$ and $0 < \Gamma(h-\alpha_h+1) \leq 1$, that is $1 - \Gamma(h-\alpha_h+1) \geq 0$.

Next we argue as in (7.82)-(7.83):

Hence we further have that

$$A(0,\lambda)=\frac{\rho_N}{\Gamma(h-\alpha_h+1)}\{(1-\lambda)-\Gamma(h-\alpha_h+1)\}\geq 0, \tag{7.88}$$

when $0\leq\lambda\leq 1-\Gamma(h-\alpha_h+1)$, proving in that case $L^*(Q_{N+n})(0)\geq 0$, given $L^*f(0)\geq 0$.

Similarly we observe that

$$A(1,\lambda)=\frac{\rho_N}{\Gamma(h-\alpha_h+1)}\{\lambda-\Gamma(h-\alpha_h+1)\}\geq 0, \tag{7.89}$$

when $\Gamma(h-\alpha_h+1)\leq\lambda\leq 1$, proving also $L^*(Q_{N+n})(1)\geq 0$ in this case, given that $L^*f(1)\geq 0$.

Clearly we have

$$A(0,0)=\frac{\rho_N}{\Gamma(h-\alpha_h+1)}\{1-\Gamma(h-\alpha_h+1)\}\geq 0, \tag{7.90}$$

proving $L^*(Q_{N+n})(0)\geq 0$, with $\lambda=0$, given $L^*f(0)\geq 0$, and

$$A(1,1)=\frac{\rho_N}{\Gamma(h-\alpha_h+1)}\{1-\Gamma(h-\alpha_h+1)\}\geq 0, \tag{7.91}$$

proving $L^*(Q_{N+n})(1)\geq 0$, given $L^*f(1)\geq 0$, with $\lambda=1$.

We see also that

$$A(x,0)=\frac{\rho_N}{\Gamma(h-\alpha_h+1)}\left\{(1-x)^{h-\alpha_h}-\Gamma(h-\alpha_h+1)\right\}\geq 0, \tag{7.92}$$

given $(1-x)\geq\Gamma(h-\alpha_h+1)^{\frac{1}{h-\alpha_h}}$, equivalently, given that $x\leq 1-\Gamma(h-\alpha_h+1)^{\frac{1}{h-\alpha_h}}$, with $h>\alpha_h$ and $x\in(0,1)$.

In that case $L^*(Q_{N+n})(x)\geq 0$, with $\lambda=0$, and h even, given there $L^*f(x)\geq 0$. At last we observe that

$$A(x,1)=\frac{\rho_N}{\Gamma(h-\alpha_h+1)}\{x^{h-\alpha_h}-\Gamma(h-\alpha_h+1)\}\geq 0, \tag{7.93}$$

given that $x\geq\Gamma(h-\alpha_h+1)^{\frac{1}{h-\alpha_h}}$, with $h>\alpha_h$ and $x\in(0,1)$.

In that case $L^*(Q_{N+n})(x)\geq 0$, with $\lambda=1$, given there $L^*f(x)\geq 0$.

II. Suppose, throughout $[0,1]$, $\alpha_h(x)\leq\overline{\beta}<0$. Call now

$$Q_{N+n}(x):=P_{N+n}(f)(x)-\rho_N\frac{x^h}{h!}. \tag{7.94}$$

Then by (7.59) we obtain

$$\begin{aligned}&\left\|\left(\lambda D_{*0}^{\alpha_j}+(1-\lambda)D_{1-}^{\alpha_j}\right)\left(f(x)-\rho_N\frac{x^h}{h!}\right)\right.\\&\left.-\left(\lambda D_{*0}^{\alpha_j}+(1-\lambda)D_{1-}^{\alpha_j}\right)(Q_{N+n}(x))\right\|_{\infty,[0,1]}\\&\leq\frac{T_N^{\beta,\alpha}(f)}{\Gamma(j-\alpha_j+1)(n-j)!},\quad 0\leq j\leq n.\end{aligned} \tag{7.95}$$

Again we obtain, as earlier, the inequalities (7.64), (7.65), (7.66). Furthermore, if $(\lambda, x) \in \Lambda$ and $L^* f(x) \geq 0$, we get

$$\alpha_h^{-1}(x) L^*(Q_{N+n}) = \alpha_h^{-1}(x) L^*(f(x))$$

$$-\rho_N \left\{ \lambda \frac{x^{h-\alpha_h}}{\Gamma(h-\alpha_h+1)} + (1-\lambda) \frac{(1-x)^{h-\alpha_h}}{\Gamma(h-\alpha_h+1)} \right\}$$

(when $\lambda \in [0,1)$ we assumed that h is even)

$$+\sum_{j=h}^{k} \alpha_h^{-1}(x) \alpha_j(x) \left\{ \left(\lambda D_{*0}^{\alpha_j} + (1-\lambda) D_{1-}^{\alpha_j}\right) \left[Q_{N+n}(x) - f(x) + \rho_N \frac{x^h}{h!} \right] \right\} \tag{7.96}$$

$$\overset{(7.95)}{\leq} -\rho_N \left\{ \lambda \frac{x^{h-\alpha_h}}{\Gamma(h-\alpha_h+1)} + (1-\lambda) \frac{(1-x)^{h-\alpha_h}}{\Gamma(h-\alpha_h+1)} \right\}$$

$$+\sum_{j=h}^{k} l_j \frac{T_N^{\beta,\alpha}(f)}{\Gamma(j-\alpha_j+1)(n-j)!}$$

$$= \rho_N \left\{ 1 - \left\{ \lambda \frac{x^{h-\alpha_h}}{\Gamma(h-\alpha_h+1)} + (1-\lambda) \frac{(1-x)^{h-\alpha_h}}{\Gamma(h-\alpha_h+1)} \right\} \right\}$$

$$= \frac{\rho_N}{\Gamma(h-\alpha_h+1)} \{\Gamma(h-\alpha_h+1) - \{\lambda x^{h-\alpha_h} + (1-\lambda)(1-x)^{h-\alpha_h}\}\} =: B(x,\lambda). \tag{7.97}$$

Hence over Λ we get that $B(x,\lambda) \leq 0$, thus there $L^*(Q_{N+n}) \geq 0$.

Next we argue as in (7.96)-(7.97):

Hence we further have that

$$B(0,\lambda) = \frac{\rho_N}{\Gamma(h-\alpha_h+1)} \{\Gamma(h-\alpha_h+1) - (1-\lambda)\} \leq 0, \tag{7.98}$$

when $0 \leq \lambda \leq 1 - \Gamma(h-\alpha_h+1)$, proving in that case $L^*(Q_{N+n})(0) \geq 0$, given $L^* f(0) \geq 0$.

Similarly we observe that

$$B(1,\lambda) = \frac{\rho_N}{\Gamma(h-\alpha_h+1)} \{\Gamma(h-\alpha_h+1) - \lambda\} \leq 0, \tag{7.99}$$

when $\Gamma(h-\alpha_h+1) \leq \lambda \leq 1$, proving also $L^*(Q_{N+n})(1) \geq 0$ in this case, given $L^* f(1) \geq 0$.

Clearly we have

$$B(0,0) = \frac{\rho_N}{\Gamma(h-\alpha_h+1)} \{\Gamma(h-\alpha_h+1) - 1\} \leq 0, \tag{7.100}$$

proving $L^*(Q_{N+n})(0) \geq 0$, given $L^* f(0) \geq 0$, with $\lambda = 0$.

Also it holds

$$B(1,1) = \frac{\rho_N}{\Gamma(h-\alpha_h+1)} \{\Gamma(h-\alpha_h+1) - 1\} \leq 0, \tag{7.101}$$

proving $L^*(Q_{N+n})(1) \geq 0$, given $L^* f(1) \geq 0$, with $\lambda = 1$.

We see also that

$$B(x,0)=\frac{\rho_N}{\Gamma(h-\alpha_h+1)}\left\{\Gamma(h-\alpha_h+1)-(1-x)^{h-\alpha_h}\right\}\leq 0, \quad (7.102)$$

given $(1-x)\geq\Gamma(h-\alpha_h+1)^{\frac{1}{h-\alpha_h}}$, with $h>\alpha_h$ and $x\in(0,1)$, h is even.

In that case $L^*(Q_{N+n})(x)\geq 0$, with $\lambda=0$, given there $L^*f(x)\geq 0$.

At last we observe that

$$B(x,1)=\frac{\rho_N}{\Gamma(h-\alpha_h+1)}\left\{\Gamma(h-\alpha_h+1)-x^{h-\alpha_h}\right\}\leq 0, \quad (7.103)$$

given that $x\geq\Gamma(h-\alpha_h+1)^{\frac{1}{h-\alpha_h}}$, with $h>\alpha_h$ and $x\in(0,1)$.

In that case again $L^*(Q_{N+n})(x)\geq 0$, with $\lambda=1$, given there $L^*f(x)\geq 0$. □

Corollary 7.2. *Let $\beta>0$, $\beta\notin\mathbb{N}$, $n=[\beta]\in\mathbb{N}:\beta=n+\alpha$, $0<\alpha<1$; $f\in AC^{n+1}([0,1])$, and $f^{(n+1)}\in L_\infty([0,1])$. Let $h,k\in\mathbb{Z}_+$ with $0\leq h\leq k\leq n$. Consider the numbers $\alpha_0=0<\alpha_1\leq 1<\alpha_2\leq 2<\alpha_3\leq 3<...<...<\alpha_n\leq n$. Let $\alpha_j(x)$, $j=h,h+1,...,k$ be real functions, defined and bounded on $[0,1]$, and suppose $\alpha_h(x)$ is either $\geq\overline{\alpha}>0$ or $\leq\overline{\beta}<0$ on $[0,1]$. We set*

$$l_\tau:\equiv\sup_{x\in[0,1]}\left|\alpha_h^{-1}(x)\,\alpha_\tau(x)\right|,\quad \tau=h,...,k. \quad (7.104)$$

Consider the left fractional linear differential operator

$$L_1:=\sum_{j=h}^{k}\alpha_j(x)\,D_{*0}^{\alpha_j}. \quad (7.105)$$

Here $T_N^{\beta,\alpha}(f)$, $N\in\mathbb{N}$, as in (7.18).

Then, for any $N\in\mathbb{N}$, there exists a real polynomial Q_{N+n} of degree $(N+n)$ such that

1) for $j=h+1,...,n$, it holds

$$\left\|D_{*0}^{\alpha_j}f-D_{*0}^{\alpha_j}Q_{N+n}\right\|_{\infty,[0,1]}\leq\frac{T_N^{\beta,\alpha}(f)}{\Gamma(j-\alpha_j+1)(n-j)!}, \quad (7.106)$$

2) for $j=1,...,h$, it holds

$$\left\|D_{*0}^{\alpha_j}(f)-D_{*0}^{\alpha_j}(Q_{N+n})\right\|_{\infty,[0,1]}\leq\frac{T_N^{\beta,\alpha}(f)}{(h-j)!}\left[\frac{1}{\Gamma(j-\alpha_j+1)}\right.$$

$$\left.+\left(\sum_{\tau=h}^{k}\frac{l_\tau}{\Gamma(\tau-\alpha_\tau+1)(n-\tau)!}\right)\left(\frac{\Gamma(h-j+1)}{\Gamma(h-\alpha_j+1)}\right)\right], \quad (7.107)$$

and

3)

$$\|f-Q_{N+n}\|_{\infty,[0,1]}\leq\frac{T_N^{\beta,\alpha}(f)}{h!}\left(\sum_{\tau=h}^{k}\frac{l_\tau}{\Gamma(\tau-\alpha_\tau+1)(n-\tau)!}+1\right). \quad (7.108)$$

We further have

1) Given $L_1f(1)\geq 0$, then $L_1(Q_{N+n})(1)\geq 0$.

2) Let $h>\alpha_h$, and $x\in(0,1)$ with $x\geq(\Gamma(h-\alpha_h+1))^{\frac{1}{h-\alpha_h}}$. Assume $L_1f(x)\geq 0$, then $L_1(Q_{N+n})(x)\geq 0$.

Proof. By Theorem 7.7 for $\lambda = 1$. □

We finish with

Corollary 7.3. *Let* $\beta > 0$, $\beta \notin \mathbb{N}$, $n = [\beta] \in \mathbb{N} : \beta = n + \alpha$, $0 < \alpha < 1$; $f \in AC^{n+1}([0,1])$, *and* $f^{(n+1)} \in L_\infty([0,1])$. *Let* $h, k \in \mathbb{Z}_+$ *with* $0 \le h \le k \le n$, h *is even. Consider the numbers* $\alpha_0 = 0 < \alpha_1 \le 1 < \alpha_2 \le 2 < \alpha_3 \le 3 < ... < ... < \alpha_n \le n$. *Let* $\alpha_j(x)$, $j = h, h+1, ..., k$ *be real functions, defined and bounded on* $[0,1]$, *and suppose* $\alpha_h(x)$ *is either* $\ge \overline{\alpha} > 0$ *or* $\le \overline{\beta} < 0$ *on* $[0,1]$. *We set*

$$l_\tau :\equiv \sup_{x\in[0,1]} \left|\alpha_h^{-1}(x)\,\alpha_\tau(x)\right|, \quad \tau = h, ..., k. \tag{7.109}$$

Consider the right fractional linear differential operator

$$L_2 := \sum_{j=h}^{k} \alpha_j(x)\, D_{1-}^{\alpha_j}. \tag{7.110}$$

Here $T_N^{\beta,\alpha}(f)$, $N \in \mathbb{N}$, *as in (7.18).*

Then, for any $N \in \mathbb{N}$, *there exists a real polynomial* Q_{N+n} *of degree* $(N+n)$ *such that*

1) for $j = h+1, ..., n$, *it holds*

$$\left\|D_{1-}^{\alpha_j} f - D_{1-}^{\alpha_j} Q_{N+n}\right\|_{\infty,[0,1]} \le \frac{T_N^{\beta,\alpha}(f)}{\Gamma(j-\alpha_j+1)(n-j)!}, \tag{7.111}$$

2) for $j = 1, ..., h$, *it holds*

$$\left\|D_{1-}^{\alpha_j}(f) - D_{1-}^{\alpha_j}(Q_{N+n})\right\|_{\infty,[0,1]} \le \frac{T_N^{\beta,\alpha}(f)}{(h-j)!}\left[\frac{1}{\Gamma(j-\alpha_j+1)}\right.$$

$$\left.+\left(\sum_{\tau=h}^{k}\frac{l_\tau}{\Gamma(\tau-\alpha_\tau+1)(n-\tau)!}\right)\left[\sum_{\theta=0}^{h-j}\binom{h-j}{\theta}\frac{\Gamma(h-j-\theta+1)}{\Gamma(h-\alpha_j-\theta+1)}\right]\right], \tag{7.112}$$

and

3)

$$\|f - Q_{N+n}\|_{\infty,[0,1]} \le \frac{T_N^{\beta,\alpha}(f)}{h!}\left(\sum_{\tau=h}^{k}\frac{l_\tau}{\Gamma(\tau-\alpha_\tau+1)(n-\tau)!}+1\right). \tag{7.113}$$

We further have

1) Given $L_2 f(0) \ge 0$, *then* $L_2(Q_{N+n})(0) \ge 0$.

2) Let even $h > \alpha_h$, $x \in (0,1)$ *such that* $x \le 1 - (\Gamma(h-\alpha_h+1))^{\frac{1}{h-\alpha_h}}$, *and* $L_2 f(x) \ge 0$. *Then* $L_2(Q_{N+n})(x) \ge 0$.

Proof. By Theorem 7.7, $\lambda = 0$. □

Bibliography

1. G.A. Anastassiou, *Fractional Korovkin theory*, Chaos, Solitons Fractals 42 (2009), 2080-2094.
2. G.A. Anastassiou, O. Shisha, *Monotone approximation with linear differential operators*, J. Approx. Theory 44 (1985), 391-393.
3. G.A. Anastassiou, *Complete fractional monotone approximation*, Acta Math. Univ. Comenian., accepted 2014.
4. K. Diethelm, *The Analysis of Fractional Differential Equations*, Lecture Notes in Mathematics, Vol. 2004, 1st edition, Springer, New York, Heidelberg, 2010.
5. A.M.A. El-Sayed, M. Gaber, *On the finite Caputo and finite Riesz derivatives*, Electronic Journal of Theoretical Physics, Vol. 3, No. 12 (2006), 81-95.
6. O. Shisha, *Monotone approximation*, Pacific J. Math. 15 (1965), 667-671.

Chapter 8

Lower Order Fractional Monotone Approximation Theory

Here is presented the theory of lower order fractional simultaneous monotone uniform polynomial approximation with rates using mixed lower order fractional linear differential operators. To obtain that, we use first ordinary simultaneous polynomial approximation with respect to the highest lower order right and left fractional derivatives of the function under approximation using their moduli of continuity. Then we use the total right and left fractional simultaneous polynomial approximation with rates, as well their convex combination. Based on the last and elegant analytical techniques, we derive preservation of monotonicity by mixed lower order fractional linear differential operators.

8.1 Introduction

The topic of monotone approximation started in [6] has become a major trend in approximation theory. A typical problem in this subject is: given a positive integer k, approximate a given function whose kth derivative is ≥ 0 by polynomials having this property.

In [3] the authors replaced the kth derivative with a linear differential operator of order k. We mention this motivating result.

Theorem 8.1. *Let h, k, p be integers, $0 \leq h \leq k \leq p$ and let f be a real function, $f^{(p)}$ continuous in $[-1,1]$ with first modulus of continuity $\omega_1\left(f^{(p)}, x\right)$ there. Let $a_j(x)$, $j = h, h+1, ..., k$ be real functions, defined and bounded on $[-1,1]$ and assume $a_h(x)$ is either $\geq$ some number $\alpha > 0$ or $\leq$ some number $\beta < 0$ throughout $[-1,1]$. Consider the operator*

$$L = \sum_{j=h}^{k} a_j(x) \left[\frac{d^j}{dx^j}\right] \tag{8.1}$$

and suppose, throughout $[-1,1]$,

$$L(f) \geq 0. \tag{8.2}$$

Then, for every integer $n \geq 1$, there is a real polynomial $Q_n(x)$ of degree $\leq n$ such that

$$L(Q_n) \geq 0 \text{ throughout } [-1,1] \tag{8.3}$$

and

$$\max_{-1 \leq x \leq 1} |f(x) - Q_n(x)| \leq C n^{k-p} \omega_1\left(f^{(p)}, \frac{1}{n}\right), \tag{8.4}$$

where C is independent of n or f.

The purpose of this chapter is to extend completely Theorem 8.1 to the lower order fractional level. All involved ordinary derivatives will become now fractional derivatives of lower order and even more we will have fractional simultaneous approximation.

We need

Definition 8.1. ([4], p. 50) Let $\alpha > 0$ and $\lceil \alpha \rceil = m$, ($\lceil \cdot \rceil$ ceiling of the number). Consider $f \in AC^m([0,1])$ (space of functions f with $f^{(m-1)} \in AC([0,1])$, absolutely continuous functions), $z \in [0,1]$. We define the left Caputo fractional derivative of f of order α as follows:

$$\left(D_{*z}^{\alpha} f\right)(x) = \frac{1}{\Gamma(m-\alpha)} \int_z^x (x-t)^{m-\alpha-1} f^{(m)}(t)\, dt, \tag{8.5}$$

for any $x \in [z,1]$, where Γ is the gamma function.

We set

$$D_{*z}^{0} f(x) = f(x),$$

$$D_{*z}^{m} f(x) = f^{(m)}(x), \ \forall\, x \in [z,1]. \tag{8.6}$$

Definition 8.2. ([5]) Let $\alpha > 0$ and $\lceil \alpha \rceil = m$. Consider $f \in AC^m([0,1])$, $z \in [0,1]$. We define the right Caputo fractional derivative of f of order α as follows:

$$\left(D_{z-}^{\alpha} f\right)(x) = \frac{(-1)^m}{\Gamma(m-\alpha)} \int_x^z (t-x)^{m-\alpha-1} f^{(m)}(t)\, dt, \tag{8.7}$$

for any $x \in [0,z]$.

We set

$$D_{z-}^{0} f(x) = f(x),$$

$$D_{z-}^{m} f(x) = (-1)^m f^{(m)}(x), \ \forall\, x \in [0,z]. \tag{8.8}$$

In particular we give

Definition 8.3. Let $0 < \alpha < 1$ and $f \in AC([0,1])$ (absolutely continuous functions), $z \in [0,1]$. We define the left Caputo fractional derivative of f of order α as follows:

$$\left(D_{*z}^{\alpha} f\right)(x) = \frac{1}{\Gamma(1-\alpha)} \int_z^x (x-t)^{-\alpha} f'(t)\, dt, \tag{8.9}$$

for any $x \in [z, 1]$.

We set

$$D_{*z}^{0} f(x) = f(x),$$
$$D_{*z}^{1} f(x) = f'(x), \ \forall\, x \in [z, 1]. \tag{8.10}$$

Definition 8.4. Let $0 < \alpha < 1$ and $f \in AC([0,1])$, $z \in [0,1]$. We define the right Caputo fractional derivative of f of order α as follows:

$$\left(D_{z-}^{\alpha} f\right)(x) = \frac{-1}{\Gamma(1-\alpha)} \int_x^z (t-x)^{-\alpha} f'(t)\, dt, \tag{8.11}$$

for any $x \in [0, z]$.

We set

$$D_{z-}^{0} f(x) = f(x),$$
$$D_{z-}^{1} f(x) = -f'(x), \ \forall\, x \in [0, z]. \tag{8.12}$$

Definition 8.5. Let $f \in C([0,1])$, we define the Bernstein polynomials

$$(B_N f)(t) = \sum_{k=0}^{N} f\left(\frac{k}{N}\right) \binom{N}{k} t^k (1-t)^{N-k}, \tag{8.13}$$

$\forall\, t \in [0,1]$, $N \in \mathbb{N}$, of degree N.

Our article [1] was the base to develop the general article [2]. We rely a lot on [2].

We need the following special result from [2], here is the $n = 1$ case.

Theorem 8.2. *Let $1 < \beta < 2$ and $0 < \alpha < 1$ such that $\beta = 1 + \alpha$. Let $f \in AC^2([0,1])$, and $f'' \in L_\infty([0,1])$, $N \in \mathbb{N}$. Set*

$$P_{N+1}(f)(x) := f(0) + \int_0^x B_N(f')(t)\, dt, \tag{8.14}$$

for all $0 \le x \le 1$, a polynomial of degree $(N+1)$. Set also

$$T_N^{\beta,\alpha}(f) := \frac{2^{1-\alpha}}{\Gamma(\alpha+1) N^{\frac{\alpha}{2}}} \left[\sup_{x \in [0,1]} \omega_1 \left(D_{x-}^{\beta} f, \frac{1}{2(\alpha+1) N^{\frac{1}{2}}} \right)_{[0,x]} \right.$$
$$\left. + \sup_{x \in [0,1]} \omega_1 \left(D_{*x}^{\beta} f, \frac{1}{2(\alpha+1) N^{\frac{1}{2}}} \right)_{[x,1]} \right] < \infty, \tag{8.15}$$

for every $N \in \mathbb{N}$.

Then

1) the quantity within the bracket of (8.15) is finite,

2) $P'_{N+1}(f) = B_N(f')$,

3)

$$\left\| P_{N+1}^{(i)}(f) - f^{(i)} \right\|_{\infty,[0,1]} \le T_N^{\beta,\alpha}(f), \quad i = 0, 1. \tag{8.16}$$

As $N \to \infty$ we derive with rates $P_{N+1}^{(i)}(f) \xrightarrow{u} f^{(i)}$ (uniformly), $i = 0, 1$.

We completely left fractionalize Theorem 8.2, to have

Theorem 8.3. *Here all terms and assumptions as in Theorem 8.2 and* $\alpha_j \in [0,1]$, $j \in \mathbb{Z}_+$. *Then*

$$\left\| D_{*0}^{\alpha_j}(f) - D_{*}^{\alpha_j}(P_{N+1}(f)) \right\|_{\infty,[0,1]} \leq \frac{T_N^{\beta,\alpha}(f)}{\Gamma(j^* - \alpha_j + 1)}, \tag{8.17}$$

where $j^* = \lceil \alpha_j \rceil = 0$ *or* 1.

Observe that (8.17) generalizes (8.16).

Proof. By [2], see there Theorem 9. □

We completely right fractionalize Theorem 8.2, to have

Theorem 8.4. *Here all terms and assumptions as in Theorem 8.2, and* $\alpha_j \in [0,1]$, $j \in \mathbb{Z}_+$. *Then*

$$\left\| D_{1-}^{\alpha_j}(f) - D_{1-}^{\alpha_j}(P_{N+1}(f)) \right\|_{\infty,[0,1]} \leq \frac{T_N^{\beta,\alpha}(f)}{\Gamma(j^* - \alpha_j + 1)}, \tag{8.18}$$

where $j^* = \lceil \alpha_j \rceil = 0$ *or* 1.

Observe that (8.18) generalizes (8.16).

Proof. By [2], see there Theorem 10. □

It follows the important

Corollary 8.1. *Here all as in Theorem 8.2,* $\lambda \in [0,1]$. *Then*

$$\left\| \left(\lambda D_{*0}^{\alpha_j}(f) + (1-\lambda) D_{1-}^{\alpha_j}(f)\right) - \right.$$

$$\left.\left(\lambda D_{*0}^{\alpha_j}(P_{N+1}(f)) + (1-\lambda) D_{1-}^{\alpha_j}(P_{N+1}(f))\right)\right\|_{\infty,[0,1]} \leq \frac{T_N^{\beta,\alpha}(f)}{\Gamma(j^* - \alpha_j + 1)}, \tag{8.19}$$

where $\alpha_j \in [0,1]$, $j \in \mathbb{Z}_+$; $j^* = \lceil \alpha_j \rceil = 0$ *or* 1.

Proof. By [2], see there Corollary 11. □

8.2 Main Results

Next comes our main result: the totally lower order fractional simultaneous monotone uniform approximation, using mixed fractional differential operators.

Theorem 8.5. *Let* $1 < \beta < 2$ *and* $0 < \alpha < 1 : \beta = 1 + \alpha$. *Let* $f \in AC^2([0,1])$ *with* $f'' \in L_\infty([0,1])$, $N \in \mathbb{N}$. *Here let* $k, \rho \in \mathbb{N} : 0 < k \leq \rho$, *and* $\alpha_0 = 0 < \alpha_1 < \alpha_2 < \ldots < \alpha_k < \ldots < \alpha_\rho \leq 1$. *Let* $\lambda \in [0,1]$, *and* $\alpha_j(x)$, $j = 0,1,\ldots,k$ *be real functions, defined and bounded on* $[0,1]$, *and suppose* $\alpha_0(x)$ *is either* $\geq \overline{\alpha} > 0$ *or* $\leq \overline{\beta} < 0$ *on* $[0,1]$. *We set*

$$l_\tau :\equiv \sup_{x \in [0,1]} \left| \alpha_0^{-1}(x)\, \alpha_\tau(x) \right|, \quad \tau = 0,1,\ldots,k. \tag{8.20}$$

Here $T_N^{\beta,\alpha}(f)$*,* $N \in \mathbb{N}$*, as in (8.15),* $(N \to \infty)$*. Consider the mixed fractional linear differential operator*

$$L_\lambda^* := \sum_{j=0}^{k} \alpha_j(x)\left[\lambda D_{*0}^{\alpha_j} + (1-\lambda) D_{1-}^{\alpha_j}\right]. \tag{8.21}$$

Then, for any $N \in \mathbb{N}$*, there exists a real polynomial* Q_{N+1} *of degree* $(N+1)$ *such that*

1)

$$\left\|\left(\lambda D_{*0}^{\alpha_j} + (1-\lambda) D_{1-}^{\alpha_j}\right)(f)\right.$$

$$\left.- \left(\lambda D_{*0}^{\alpha_j} + (1-\lambda) D_{1-}^{\alpha_j}\right)(Q_{N+1}(f))\right\|_{\infty,[0,1]} \le \frac{T_N^{\beta,\alpha}(f)}{\Gamma(2-\alpha_j)},\ j = 1, ..., \rho, \tag{8.22}$$

and

2)

$$\|f - Q_{N+1}(f)\|_{\infty,[0,1]} \le T_N^{\beta,\alpha}(f)\left[\sum_{\tau=1}^{k} \frac{l_\tau}{\Gamma(2-\alpha_\tau)} + 2\right]. \tag{8.23}$$

Assuming $L_\lambda^* f(x) \ge 0$*, for all* $x \in [0,1]$ *we get* $(L_\lambda^*(Q_{N+1}(f)))(x) \ge 0$ *for all* $x \in [0,1]$*.*

Proof. Here let $k, \rho \in \mathbb{N} : 0 < k \le \rho$, and $\alpha_0 = 0 < \alpha_1 < \alpha_2 < ... < \alpha_k < ... < \alpha_\rho \le 1$, that is $\lceil \alpha_0 \rceil = 0$; $\lceil \alpha_j \rceil = 1$, $j = 1, ..., \rho$. We set

$$l_\tau := \sup_{x \in [0,1]} \left|\alpha_0^{-1}(x)\,\alpha_\tau(x)\right| < \infty,\quad 0 \le \tau \le k, \tag{8.24}$$

with $l_0 = 1$, and

$$\rho_N := T_N^{\beta,\alpha}(f)\left(\sum_{\tau=1}^{k} \frac{l_\tau}{\Gamma(2-\alpha_\tau)} + 1\right). \tag{8.25}$$

I. Suppose, throughout $[0,1]$, $\alpha_0(x) \ge \overline{\alpha} > 0$. Call

$$Q_{N+1}(f)(x) := P_{N+1}(f)(x) + \rho_N, \tag{8.26}$$

where $P_{N+1}(f)(x)$ as in (8.14).

Then by (8.19) we obtain

$$\left\|\left(\lambda D_{*0}^{\alpha_j} + (1-\lambda) D_{1-}^{\alpha_j}\right)(f(x) + \rho_N)\right.$$

$$\left.- \left(\lambda D_{*0}^{\alpha_j} + (1-\lambda) D_{1-}^{\alpha_j}\right)(Q_{N+1}(f)(x))\right\|_{\infty,[0,1]} \le \frac{T_N^{\beta,\alpha}(f)}{\Gamma(2-\alpha_j)},\quad j = 1, ..., \rho. \tag{8.27}$$

And of course it holds

$$\|(f(x) + \rho_N) - Q_{N+1}(f)(x)\|_{\infty,[0,1]} \overset{(8.16)}{\le} T_N^{\beta,\alpha}(f). \tag{8.28}$$

So that we find

$$\|f - Q_{N+1}(f)\|_{\infty,[0,1]} \overset{(8.28)}{\leq} \rho_N + T_N^{\beta,\alpha}(f) \tag{8.29}$$

$$= T_N^{\beta,\alpha}(f)\left(\sum_{\tau=1}^{k}\frac{l_\tau}{\Gamma(2-\alpha_j)} + 1\right) + T_N^{\beta,\alpha}(f) = T_N^{\beta,\alpha}(f)\left[\sum_{\tau=1}^{k}\frac{l_\tau}{\Gamma(2-\alpha_j)} + 2\right],$$

proving (8.23).

From (8.27) and (8.9), (8.11), we get

$$\left\|\left(\lambda D_{*0}^{\alpha_j} + (1-\lambda) D_{1-}^{\alpha_j}\right)(f(x))\right.$$

$$\left. - \left(\lambda D_{*0}^{\alpha_j} + (1-\lambda) D_{1-}^{\alpha_j}\right)(Q_{N+1}(f))(x)\right\|_{\infty,[0,1]} \leq \frac{T_N^{\beta,\alpha}(f)}{\Gamma(2-\alpha_j)}, \quad j = 1, ..., \rho, \tag{8.30}$$

proving (8.22).

Next we use the assumption $L_\lambda^* f(x) \geq 0$, all $x \in [0,1]$, to get

$$\alpha_0^{-1}(x) L_\lambda^*(Q_{N+1}(f))(x) = \alpha_0^{-1}(x) L_\lambda^* f(x) + \rho_N$$

$$+ \sum_{j=0}^{k} \alpha_0^{-1}(x)\alpha_j(x)\left\{\left(\lambda D_{*0}^{\alpha_j} + (1-\lambda) D_{1-}^{\alpha_j}\right)[Q_{N+1}(f)(x) - f(x) - \rho_N]\right\}$$

$$\overset{((8.27),(8.28))}{\geq} \rho_N - \left(\sum_{\tau=1}^{k}\frac{l_\tau}{\Gamma(2-\alpha_j)} + 1\right) T_N^{\beta,\alpha}(f) = \rho_N - \rho_N = 0. \tag{8.31}$$

Hence $L_\lambda^*(Q_{N+1}(f))(x) \geq 0$, all $x \in [0,1]$.

II. Suppose, throughout $[0,1]$, $\alpha_0(x) \leq \overline{\beta} < 0$. Call

$$Q_{N+1}(f)(x) := P_{N+1}(f)(x) - \rho_N, \tag{8.32}$$

where $P_{N+1}(f)(x)$ as in (8.14).

Then by (8.19) we obtain

$$\left\|\left(\lambda D_{*0}^{\alpha_j} + (1-\lambda) D_{1-}^{\alpha_j}\right)(f(x) - \rho_N)\right.$$

$$\left. - \left(\lambda D_{*0}^{\alpha_j} + (1-\lambda) D_{1-}^{\alpha_j}\right)(Q_{N+1}(f)(x))\right\|_{\infty,[0,1]} \leq \frac{T_N^{\beta,\alpha}(f)}{\Gamma(2-\alpha_j)}, \quad j = 1, ..., \rho. \tag{8.33}$$

And of course it holds

$$\|(f(x) - \rho_N) - Q_{N+1}(f)(x)\|_{\infty,[0,1]} \overset{(8.16)}{\leq} T_N^{\beta,\alpha}(f). \tag{8.34}$$

Similarly we obtain again (8.22) and (8.23).

Next we use the assumption $L_\lambda^* f(x) \geq 0$, all $x \in [0,1]$, to get

$$\alpha_0^{-1}(x) L_\lambda^*(Q_{N+1}(f))(x) = \alpha_0^{-1}(x) L_\lambda^* f(x) - \rho_N +$$

$$\sum_{j=0}^{k} \alpha_0^{-1}(x)\alpha_j(x)\left\{\left(\lambda D_{*0}^{\alpha_j} + (1-\lambda) D_{1-}^{\alpha_j}\right)[Q_{N+1}(f)(x) - f(x) + \rho_N]\right\}$$

$$\overset{((8.33),(8.34))}{\leq} -\rho_N + \left(\sum_{\tau=1}^{k}\frac{l_\tau}{\Gamma(2-\alpha_j)} + 1\right) T_N^{\beta,\alpha}(f) = -\rho_N + \rho_N = 0. \tag{8.35}$$

Hence $L_\lambda^*(Q_{N+1}(f))(x) \geq 0$, for any $x \in [0,1]$. $\square$

Corollary 8.2. *(to Theorem 8.5, $\lambda = 1$ case) Let $L_1^* := \sum_{j=0}^{k} \alpha_j(x) D_{*0}^{\alpha_j}$.*

Then, for any $N \in \mathbb{N}$, there exists a real polynomial Q_{N+1} of degree $(N+1)$ such that

1)

$$\left\| D_{*0}^{\alpha_j}(f) - D_{*0}^{\alpha_j}(Q_{N+1}(f)) \right\|_{\infty,[0,1]} \leq \frac{T_N^{\beta,\alpha}(f)}{\Gamma(2-\alpha_j)}, \quad j = 1, ..., \rho, \tag{8.36}$$

2) inequality (8.23) is again valid.

Assuming $L_1^* f(x) \geq 0$, for all $x \in [0,1]$, we get $(L_1^*(Q_{N+1}(f)))(x) \geq 0$ for all $x \in [0,1]$.

Corollary 8.3. *(to Theorem 8.5, $\lambda = 0$ case) Let $L_0^* := \sum_{j=0}^{k} \alpha_j(x) D_{1-}^{\alpha_j}$.*

Then, for any $N \in \mathbb{N}$, there exists a real polynomial Q_{N+1} of degree $(N+1)$ such that

1)

$$\left\| D_{1-}^{\alpha_j} f - D_{1-}^{\alpha_j}(Q_{N+1}(f)) \right\|_{\infty,[0,1]} \leq \frac{T_N^{\beta,\alpha}(f)}{\Gamma(2-\alpha_j)}, \quad j = 1, ..., \rho, \tag{8.37}$$

2) inequality (8.23) is again valid.

Assuming $L_0^* f(x) \geq 0$, we get $L_0^*(Q_{N+1}(f))(x) \geq 0$ for any $x \in [0,1]$.

Finally we give

Corollary 8.4. *(to Theorem 8.5, $\lambda = \frac{1}{2}$ case) Let $L_{\frac{1}{2}}^* := \sum_{j=0}^{k} \alpha_j(x) [\frac{D_{*0}^{\alpha_j} + D_{1-}^{\alpha_j}}{2}]$.*

Then, for any $N \in \mathbb{N}$, there exists a real polynomial Q_{N+1} of degree $(N+1)$ such that

1)

$$\left\| (D_{*0}^{\alpha_j} + D_{1-}^{\alpha_j})(f) - (D_{*0}^{\alpha_j} + D_{1-}^{\alpha_j})(Q_{N+1}(f)) \right\|_{\infty,[0,1]} \leq \frac{2T_N^{\beta,\alpha}(f)}{\Gamma(2-\alpha_j)}, \quad j = 1, ..., \rho, \tag{8.38}$$

2) inequality (8.23) is again valid.

Assuming $L_{\frac{1}{2}}^* f(x) \geq 0$, we get $L_{\frac{1}{2}}^*(Q_{N+1}(f))(x) \geq 0$ for any $x \in [0,1]$.

Bibliography

1. G.A. Anastassiou, *Fractional Korovkin theory*, Chaos, Solitons Fractals 42 (2009), 2080-2094.
2. G.A. Anastassiou, *Complete fractional monotone approximation*, Acta Math. Univ. Comenian., accepted 2014.
3. G.A. Anastassiou, O. Shisha, *Monotone approximation with linear differential operators*, J. Approx. Theory 44 (1985), 391-393.
4. K. Diethelm, *The Analysis of Fractional Differential Equations*, Lecture Notes in Mathematics, Vol. 2004, 1st edition, Springer, New York, Heidelberg, 2010.
5. A.M.A. El-Sayed, M. Gaber, *On the finite Caputo and finite Riesz derivatives*, Electron. J. Theoret. Phys. 3(12) (2006), 81-95.
6. O. Shisha, *Monotone approximation*, Pacific J. Math. 15 (1965), 667-671.

Chapter 9

Approximation Theory by Discrete Singular Operators

Here we study basic approximation properties with rates of our discrete versions of Picard, Gauss-Weierstrass, Poisson-Cauchy singular operators and of two other discrete operators. We prove uniform convergence of these operators to the unit operator. Also all these operators fulfill the global smoothness preservation property. The discussed operators act on the space of uniformly continuous functions over the real line. It follows [2].

9.1 Preliminaries

Let $f : \mathbb{R} \to \mathbb{R}$ be a function which is uniformly continuous ($f \in C_U(\mathbb{R})$). Following [3], p. 40-41, we define the first modulus of continuity,

$$\omega_1(f,t) := \sup_{\substack{x,y \in \mathbb{R} \\ |x-y| \leq t}} |f(x) - f(y)|, \quad t \geq 0. \tag{9.1}$$

The function ω_1 is continuous at $t = 0$ if and only if f is uniformly continuous on $\mathbb{R}$. So that here $\omega_1(f,t) \to \omega_1(f,0) = 0$, as $t \to 0$. It also holds

$$\omega_1(f, \lambda t) \leq (\lambda + 1)\,\omega_1(f,t), \quad \lambda \geq 0. \tag{9.2}$$

Clearly $\omega_1(f,t)$ is finite for each $t \geq 0$.

In [1] we studied extensively the convergence to the unit operator of various integral singular operators. Here we define the discrete analogs of these operators next, and we study their uniform convergence to the unit operator with rates.

Let $0 < \xi \leq 1$, such that $\xi \to 0+$, $x \in \mathbb{R}$; $\frac{1}{\xi} \geq 1$.

i) We define the discrete Picard operators:

$$(P_\xi^* f)(x) := \frac{\sum_{\nu=-\infty}^{\infty} f(x+\nu)\, e^{-\frac{|\nu|}{\xi}}}{\sum_{\nu=-\infty}^{\infty} e^{-\frac{|\nu|}{\xi}}}. \tag{9.3}$$

ii) We define the discrete Gauss-Weierstrass operators:

$$(W_\xi^* f)(x) := \frac{\sum_{\nu=-\infty}^{\infty} f(x+\nu)\, e^{-\frac{\nu^2}{\xi}}}{\sum_{\nu=-\infty}^{\infty} e^{-\frac{\nu^2}{\xi}}}. \tag{9.4}$$

iii) We define the general discrete Poisson-Cauchy operators:
let $\alpha \in \mathbb{N}$, $\beta > \frac{1}{\alpha}$;

$$\left(M_{\xi}^{*} f\right)(x) := \frac{\sum_{\nu=-\infty}^{\infty} f(x+\nu)\left(\nu^{2\alpha}+\xi^{2\alpha}\right)^{-\beta}}{\sum_{\nu=-\infty}^{\infty}\left(\nu^{2\alpha}+\xi^{2\alpha}\right)^{-\beta}}. \tag{9.5}$$

iv) We define the basic discrete convolution operators:
let $\varphi : \mathbb{R} \to \mathbb{R}$, with $\|\varphi\|_{\infty} := \sup_{x\in\mathbb{R}} |\varphi(x)| \leq K$, $K > 0$, $\beta \in \mathbb{N} - \{1\}$;

$$\left(\theta_{\xi}^{*} f\right)(x) := \frac{f(x) + \sum_{\nu\in\mathbb{Z}-\{0\}} f(x+\nu)\left(\frac{\varphi\left(\frac{\nu}{\xi}\right)}{\frac{\nu}{\xi}}\right)^{2\beta}}{1 + \sum_{\nu\in\mathbb{Z}-\{0\}}\left(\frac{\varphi\left(\frac{\nu}{\xi}\right)}{\frac{\nu}{\xi}}\right)^{2\beta}}. \tag{9.6}$$

v) We define the general discrete convolution operators:
let $\varphi : \mathbb{R} \to \mathbb{R}_{+}$ with $\varphi(x) \leq Ax^{2\beta}$, $\forall\, x \in \mathbb{R}$, $\beta \in \mathbb{N} - \{1\}$, $A > 0$;

$$\left(T_{\xi}^{*} f\right)(x) := \frac{f(x) + \sum_{\nu\in\mathbb{Z}-\{0\}} f(x+\nu)\frac{\varphi\left(\frac{\nu}{\xi}\right)}{\left(\frac{\nu}{\xi}\right)^{4\beta}}}{1 + \sum_{\nu\in\mathbb{Z}-\{0\}}\frac{\varphi\left(\frac{\nu}{\xi}\right)}{\left(\frac{\nu}{\xi}\right)^{4\beta}}}. \tag{9.7}$$

The above operators, as we will see, are well defined and are linear, positive, and bounded when $\|f\|_{\infty} := \sup_{x\in\mathbb{R}} |f(x)| < \infty$. Furthermore

$$P_{\xi}^{*}(1) = W_{\xi}^{*}(1) = M_{\xi}^{*}(1) = \theta_{\xi}^{*}(1) = T_{\xi}^{*}(1) = 1, \tag{9.8}$$

with

$$\left\|P_{\xi}^{*}\right\| = \left\|W_{\xi}^{*}\right\| = \left\|M_{\xi}^{*}\right\| = \left\|\theta_{\xi}^{*}\right\| = \left\|T_{\xi}^{*}\right\| = 1, \tag{9.9}$$

on continuous bounded functions.

In this chapter we are motivated by [4].

9.2 Main Results

All here as in Preliminaries earlier. We start with the basic approximation properties of discrete Picard operators.

We present

Theorem 9.1. *It holds*

$$\left\|P_{\xi}^{*} f - f\right\|_{\infty} \leq \left[\frac{1 + 2e^{-\frac{1}{\xi}}\left(2\xi + 2 + \frac{1}{\xi}\right)}{1 + 2\xi e^{-\frac{1}{\xi}}}\right] \omega_1(f, \xi). \tag{9.10}$$

The constant in the right hand side of (9.10) converges to 1 *as* $\xi \to 0+$. *So that* $P_{\xi}^{*} \to I$ *(unit operator), uniformly with rates, as* $\xi \to 0+$.

Proof. We will use a lot

$$\sum_{\nu=1}^{\infty} \frac{1}{\nu^2} = \frac{\pi^2}{6} \quad \text{(Euler, 1741)}.$$

We see that

$$\sum_{\nu=-1}^{-\infty} e^{-\frac{|\nu|}{\xi}} = \sum_{\nu=1}^{\infty} e^{-\frac{\nu}{\xi}} < \sum_{\nu=1}^{\infty} \frac{1}{\nu^2} = \frac{\pi^2}{6},$$

it converges.

Thus

$$\sum_{\nu=-\infty}^{\infty} e^{-\frac{|\nu|}{\xi}} = 2\sum_{\nu=1}^{\infty} e^{-\frac{\nu}{\xi}} + 1 < \frac{\pi^2}{3} + 1. \tag{9.11}$$

Using [5] we obtain

$$\sum_{\nu=1}^{\infty} e^{-\frac{\nu}{\xi}} - e^{-\frac{1}{\xi}} \le \int_1^{\infty} e^{-\frac{\nu}{\xi}} d\nu \le \sum_{\nu=1}^{\infty} e^{-\frac{\nu}{\xi}}. \tag{9.12}$$

Hence

$$2\int_1^{\infty} e^{-\frac{\nu}{\xi}} d\nu + 1 \le 2\sum_{\nu=1}^{\infty} e^{-\frac{\nu}{\xi}} + 1 = \sum_{\nu=-\infty}^{\infty} e^{-\frac{|\nu|}{\xi}}. \tag{9.13}$$

Thus

$$0 < \frac{1}{\sum_{\nu=-\infty}^{\infty} e^{-\frac{|\nu|}{\xi}}} \le \frac{1}{2\int_1^{\infty} e^{-\frac{\nu}{\xi}} d\nu + 1} = \frac{1}{2\xi e^{-\frac{1}{\xi}} + 1} \to 1, \text{ as } \xi \to 0+. \tag{9.14}$$

We need to prove that $g(\nu) = \nu e^{-\frac{\nu}{\xi}}$ is decreasing for $\nu \ge 1$. Indeed we have that $g'(\nu) = e^{-\frac{\nu}{\xi}} \left(1 - \frac{\nu}{\xi}\right) \le 0$, by $\xi \le 1 \le \nu$.

So that, again by [5], we get that

$$1 + 2\sum_{\nu=1}^{\infty} \left(1 + \frac{\nu}{\xi}\right) e^{-\frac{\nu}{\xi}} \le 1 + 2\left[\int_1^{\infty} \left(1 + \frac{\nu}{\xi}\right) e^{-\frac{\nu}{\xi}} d\nu + \left(1 + \frac{1}{\xi}\right) e^{-\frac{1}{\xi}}\right] =: (*). \tag{9.15}$$

Using integration by parts we have

$$\int_{\frac{1}{\xi}}^{\infty} x e^{-x} dx = -e^{-x}(x+1)\Big|_{\frac{1}{\xi}}^{\infty} = e^{-\frac{1}{\xi}} \left(\frac{1}{\xi} + 1\right). \tag{9.16}$$

Hence we get

$$\int_1^{\infty} \left(1 + \frac{\nu}{\xi}\right) e^{-\frac{\nu}{\xi}} d\nu = \int_1^{\infty} e^{-\frac{\nu}{\xi}} d\nu + \int_1^{\infty} \frac{\nu}{\xi} e^{-\frac{\nu}{\xi}} d\nu$$

$$= \xi e^{-\frac{1}{\xi}} + \xi \int_{\frac{1}{\xi}}^{\infty} x e^{-x} dx = e^{-\frac{1}{\xi}} (2\xi + 1). \tag{9.17}$$

Therefore

$$(*) = 1 + 2\left[e^{-\frac{1}{\xi}}(2\xi+1) + \left(1+\frac{1}{\xi}\right)e^{-\frac{1}{\xi}}\right] = 1 + 2e^{-\frac{1}{\xi}}\left(2\xi+2+\frac{1}{\xi}\right). \tag{9.18}$$

Consequently we have found that

$$\sum_{\nu=-\infty}^{\infty}\left(1+\frac{|\nu|}{\xi}\right)e^{-\frac{|\nu|}{\xi}} = 1 + 2\sum_{\nu=1}^{\infty}\left(1+\frac{\nu}{\xi}\right)e^{-\frac{\nu}{\xi}}$$

$$\leq 1 + 2e^{-\frac{1}{\xi}}\left(2\xi+2+\frac{1}{\xi}\right)\text{ (finite)} \to 1, \text{ as } \xi \to 0+. \tag{9.19}$$

Finally we observe

$$\left(P_{\xi}^{*}f\right)(x) - f(x) = \frac{\sum_{\nu=-\infty}^{\infty}\left(f(x+\nu) - f(x)\right)e^{-\frac{|\nu|}{\xi}}}{\sum_{\nu=-\infty}^{\infty}e^{-\frac{|\nu|}{\xi}}}. \tag{9.20}$$

So that

$$\left|\left(P_{\xi}^{*}f\right)(x) - f(x)\right| \leq \frac{\sum_{\nu=-\infty}^{\infty}\left|f(x+\nu) - f(x)\right|e^{-\frac{|\nu|}{\xi}}}{\sum_{\nu=-\infty}^{\infty}e^{-\frac{|\nu|}{\xi}}} \tag{9.21}$$

$$\leq \frac{\sum_{\nu=-\infty}^{\infty}\omega_1(f,|\nu|)e^{-\frac{|\nu|}{\xi}}}{\sum_{\nu=-\infty}^{\infty}e^{-\frac{|\nu|}{\xi}}} = \frac{\sum_{\nu=-\infty}^{\infty}\omega_1\left(f,\xi\frac{|\nu|}{\xi}\right)e^{-\frac{|\nu|}{\xi}}}{\sum_{\nu=-\infty}^{\infty}e^{-\frac{|\nu|}{\xi}}}$$

(by (9.2))

$$\leq \omega_1(f,\xi)\left(\frac{\sum_{\nu=-\infty}^{\infty}\left(1+\frac{|\nu|}{\xi}\right)e^{-\frac{|\nu|}{\xi}}}{\sum_{\nu=-\infty}^{\infty}e^{-\frac{|\nu|}{\xi}}}\right) \tag{9.22}$$

(by (9.14), (9.19))

$$\leq \omega_1(f,\xi)\frac{\left(1+2e^{-\frac{1}{\xi}}\left(2\xi+2+\frac{1}{\xi}\right)\right)}{\left(2\xi e^{-\frac{1}{\xi}}+1\right)}. \tag{9.23}$$

We notice that

$$\frac{\left(1+2e^{-\frac{1}{\xi}}\left(2\xi+2+\frac{1}{\xi}\right)\right)}{\left(2\xi e^{-\frac{1}{\xi}}+1\right)} \to 1, \quad \text{as } \xi \to 0+.$$

We have proved

$$\left|\left(P_{\xi}^{*}f\right)(x) - f(x)\right| \leq \left[\frac{1+2e^{-\frac{1}{\xi}}\left(2\xi+2+\frac{1}{\xi}\right)}{1+2\xi e^{-\frac{1}{\xi}}}\right]\omega_1(f,\xi), \tag{9.24}$$

$\forall\, x \in \mathbb{R}$.

The proof now is completed. □

Next we prove preservation of global smoothness of P_ξ^*.

Theorem 9.2. *It holds*

$$\omega_1\left(P_\xi^* f, \delta\right) \leq \omega_1(f, \delta), \quad \forall\, \delta > 0. \tag{9.25}$$

Inequality (9.25) is sharp, namely it is attained by $f(x) = identity(x) = x$.

Proof. We see that

$$\left(P_\xi^* f\right)(x) - \left(P_\xi^* f\right)(y) = \frac{\sum_{\nu=-\infty}^{\infty} \left(f(x+\nu) - f(y+\nu)\right) e^{-\frac{|\nu|}{\xi}}}{\sum_{\nu=-\infty}^{\infty} e^{-\frac{|\nu|}{\xi}}}. \tag{9.26}$$

Hence

$$\left|\left(P_\xi^* f\right)(x) - \left(P_\xi^* f\right)(y)\right| \leq \frac{\sum_{\nu=-\infty}^{\infty} \left|f(x+\nu) - f(y+\nu)\right| e^{-\frac{|\nu|}{\xi}}}{\sum_{\nu=-\infty}^{\infty} e^{-\frac{|\nu|}{\xi}}}$$

$$\leq \frac{\sum_{\nu=-\infty}^{\infty} \omega_1(f, |x-y|)\, e^{-\frac{|\nu|}{\xi}}}{\sum_{\nu=-\infty}^{\infty} e^{-\frac{|\nu|}{\xi}}} = \omega_1(f, |x-y|). \tag{9.27}$$

So that for any $x, y \in \mathbb{R} : |x-y| < \delta$ we get (9.25).

If $f = id$, then trivially we get

$$\left(P_\xi^* id\right)(x) - \left(P_\xi^* id\right)(y) = x - y = id(x) - id(y), \tag{9.28}$$

thus (9.25) is attained. □

Next we study the approximation properties of discrete Gauss-Weierstrass operators.

Theorem 9.3. *Let $f \in C_U(\mathbb{R})$, $0 < \xi \leq 1$. Then*

$$\left\|W_\xi^* f - f\right\|_\infty \leq C(\xi)\, \omega_1\left(f, \sqrt{\xi}\right), \tag{9.29}$$

where

$$C(\xi) := \left[1 + \left(\frac{e^{-\frac{1}{\xi}}\left(\sqrt{\xi} + 2 + \frac{2}{\sqrt{\xi}}\right)}{\sqrt{\pi\xi}\left(1 - \operatorname{erf}\left(\frac{1}{\sqrt{\xi}}\right)\right) + 1}\right)\right]. \tag{9.30}$$

We have $\lim_{\xi \to 0+} C(\xi) = 1$, and by $\lim_{\xi \to 0+} \omega_1\left(f, \sqrt{\xi}\right) = 0$, we get $W_\xi^* \to I$ uniformly with rates, as $\xi \to 0+$.

Proof. We notice easily that

$$\sum_{\nu=-1}^{-\infty} e^{-\frac{\nu^2}{\xi}} = \sum_{\nu=1}^{\infty} e^{-\frac{\nu^2}{\xi}} < \sum_{\nu=1}^{\infty} \frac{1}{\nu^2} = \frac{\pi^2}{6} < \infty. \tag{9.31}$$

So we can write

$$\sum_{\nu=-\infty}^{\infty} e^{-\frac{\nu^2}{\xi}} = 2\sum_{\nu=1}^{\infty} e^{-\frac{\nu^2}{\xi}} + 1 < \frac{\pi^2}{3} + 1. \tag{9.32}$$

Since $e^{-\frac{\nu^2}{\xi}}$ is positive, continuous and decreasing, by [5], we get

$$\sum_{\nu=1}^{\infty} e^{-\frac{\nu^2}{\xi}} - e^{-\frac{1}{\xi}} \leq \int_1^{\infty} e^{-\frac{\nu^2}{\xi}} d\nu \leq \sum_{\nu=1}^{\infty} e^{-\frac{\nu^2}{\xi}}. \tag{9.33}$$

So that

$$2\int_1^{\infty} e^{-\frac{\nu^2}{\xi}} d\nu + 1 \leq 2\sum_{\nu=1}^{\infty} e^{-\frac{\nu^2}{\xi}} + 1 = \sum_{\nu=-\infty}^{\infty} e^{-\frac{\nu^2}{\xi}}, \tag{9.34}$$

and

$$0 < \frac{1}{\sum_{\nu=-\infty}^{\infty} e^{-\frac{\nu^2}{\xi}}} \leq \frac{1}{2\int_1^{\infty} e^{-\frac{\nu^2}{\xi}} d\nu + 1}. \tag{9.35}$$

We know that $\int_0^{\infty} e^{-t^2} dt = \frac{\sqrt{\pi}}{2}$, and $\operatorname{erf}(x) := \frac{2}{\sqrt{\pi}} \int_0^x e^{-t^2} dt$, with $\operatorname{erf}(\infty) = 1$.

Hence

$$2\int_1^{\infty} e^{-\frac{\nu^2}{\xi}} d\nu + 1 = 2\sqrt{\xi} \int_1^{\infty} e^{-\left(\frac{\nu}{\sqrt{\xi}}\right)^2} d\left(\frac{\nu}{\sqrt{\xi}}\right) + 1 \tag{9.36}$$

$$= 2\sqrt{\xi} \int_{\frac{1}{\sqrt{\xi}}}^{\infty} e^{-\theta^2} d\theta + 1 = 2\sqrt{\xi}\left[\int_0^{\infty} e^{-\theta^2} d\theta - \int_0^{\frac{1}{\sqrt{\xi}}} e^{-\theta^2} d\theta\right] + 1$$

$$= 2\sqrt{\xi}\left[\frac{\sqrt{\pi}}{2} - \frac{\sqrt{\pi}}{2} \operatorname{erf}\left(\frac{1}{\sqrt{\xi}}\right)\right] + 1 = \sqrt{\pi\xi}\left(1 - \operatorname{erf}\left(\frac{1}{\sqrt{\xi}}\right)\right) + 1. \tag{9.37}$$

Therefore

$$2\int_1^{\infty} e^{-\frac{\nu^2}{\xi}} d\nu + 1 = \sqrt{\pi\xi}\left(1 - \operatorname{erf}\left(\frac{1}{\sqrt{\xi}}\right)\right) + 1 \to 1, \text{ as } \xi \to 0+. \tag{9.38}$$

So we got that

$$0 < \frac{1}{\sum_{\nu=-\infty}^{\infty} e^{-\frac{\nu^2}{\xi}}} \leq \frac{1}{\sqrt{\pi\xi}\left(1 - \operatorname{erf}\left(\frac{1}{\sqrt{\xi}}\right)\right) + 1} \to 1, \text{ as } \xi \to 0+. \tag{9.39}$$

Next we prove that $g(\nu) = \nu e^{-\frac{\nu^2}{\xi}}$ is decreasing for $\nu \geq 1$. Indeed we have $g'(\nu) = e^{-\frac{\nu^2}{\xi}}\left(1 - \frac{2\nu^2}{\xi}\right) \leq 0$, iff $1 - \frac{2\nu^2}{\xi} \leq 0$, iff $\xi \leq 2\nu^2$, which is true.

So that we have (by [5])

$$\sum_{\nu=1}^{\infty}\left(1 + \frac{\nu}{\sqrt{\xi}}\right) e^{-\frac{\nu^2}{\xi}} \leq \int_1^{\infty}\left(1 + \frac{\nu}{\sqrt{\xi}}\right) e^{-\frac{\nu^2}{\xi}} d\nu + \left(1 + \frac{1}{\sqrt{\xi}}\right) e^{-\frac{1}{\xi}} \tag{9.40}$$

$$= \int_1^{\infty} e^{-\frac{\nu^2}{\xi}} d\nu + \int_1^{\infty} \frac{\nu}{\sqrt{\xi}} e^{-\frac{\nu^2}{\xi}} d\nu + e^{-\frac{1}{\xi}} + \frac{e^{-\frac{1}{\xi}}}{\sqrt{\xi}}$$

$$= \frac{\sqrt{\pi\xi}}{2}\left(1 - \operatorname{erf}\left(\frac{1}{\sqrt{\xi}}\right)\right) + \frac{\sqrt{\xi}}{2}\int_1^{\infty} e^{-\frac{\nu^2}{\xi}} d\left(\frac{\nu^2}{\xi}\right) + e^{-\frac{1}{\xi}} + \frac{e^{-\frac{1}{\xi}}}{\sqrt{\xi}}$$

$$= \frac{\sqrt{\pi\xi}}{2}\left(1-\text{erf}\left(\frac{1}{\sqrt{\xi}}\right)\right)+\frac{\sqrt{\xi}}{2}\int_{\frac{1}{\xi}}^{\infty}e^{-x}dx+e^{-\frac{1}{\xi}}+\frac{e^{-\frac{1}{\xi}}}{\sqrt{\xi}}$$

$$= \frac{\sqrt{\pi\xi}}{2}\left(1-\text{erf}\left(\frac{1}{\sqrt{\xi}}\right)\right)+\frac{\sqrt{\xi}}{2}e^{-\frac{1}{\xi}}+e^{-\frac{1}{\xi}}+\frac{e^{-\frac{1}{\xi}}}{\sqrt{\xi}}. \tag{9.41}$$

That is

$$\sum_{\nu=1}^{\infty}\left(1+\frac{\nu}{\sqrt{\xi}}\right)e^{-\frac{\nu^2}{\xi}} \leq \frac{\sqrt{\pi\xi}}{2}\left(1-\text{erf}\left(\frac{1}{\sqrt{\xi}}\right)\right)+e^{-\frac{1}{\xi}}\left(\frac{\sqrt{\xi}}{2}+1+\frac{1}{\sqrt{\xi}}\right) \tag{9.42}$$

(finite) $\rightarrow 0$, as $\xi \rightarrow 0+$.

Since

$$\sum_{\nu=-\infty}^{\infty}\left(1+\frac{|\nu|}{\sqrt{\xi}}\right)e^{-\frac{\nu^2}{\xi}} = 2\sum_{\nu=1}^{\infty}\left(1+\frac{\nu}{\sqrt{\xi}}\right)e^{-\frac{\nu^2}{\xi}}+1<\infty, \tag{9.43}$$

we find

$$\sum_{\nu=-\infty}^{\infty}\left(1+\frac{|\nu|}{\sqrt{\xi}}\right)e^{-\frac{\nu^2}{\xi}} \leq \sqrt{\pi\xi}\left(1-\text{erf}\left(\frac{1}{\sqrt{\xi}}\right)\right)+e^{-\frac{1}{\xi}}\left(\sqrt{\xi}+2+\frac{2}{\sqrt{\xi}}\right)+1 \tag{9.44}$$

(is finite) $\rightarrow 1$, as $\xi \rightarrow 0+$.

Next we observe that

$$\left(W_{\xi}^{*}f\right)(x)-f(x) = \frac{\sum_{\nu=-\infty}^{\infty}\left(f(x+\nu)-f(x)\right)e^{-\frac{\nu^2}{\xi}}}{\sum_{\nu=-\infty}^{\infty}e^{-\frac{\nu^2}{\xi}}}. \tag{9.45}$$

Thus

$$\left|\left(W_{\xi}^{*}f\right)(x)-f(x)\right| \leq \frac{\sum_{\nu=-\infty}^{\infty}\left|f(x+\nu)-f(x)\right|e^{-\frac{\nu^2}{\xi}}}{\sum_{\nu=-\infty}^{\infty}e^{-\frac{\nu^2}{\xi}}} \tag{9.46}$$

$$\leq \frac{\sum_{\nu=-\infty}^{\infty}\omega_1(f,|\nu|)e^{-\frac{\nu^2}{\xi}}}{\sum_{\nu=-\infty}^{\infty}e^{-\frac{\nu^2}{\xi}}} = \frac{\sum_{\nu=-\infty}^{\infty}\omega_1\left(f,\sqrt{\xi}\frac{|\nu|}{\sqrt{\xi}}\right)e^{-\frac{\nu^2}{\xi}}}{\sum_{\nu=-\infty}^{\infty}e^{-\frac{\nu^2}{\xi}}} \tag{9.47}$$

$$\leq \frac{\omega_1\left(f,\sqrt{\xi}\right)\sum_{\nu=-\infty}^{\infty}\left(1+\frac{|\nu|}{\sqrt{\xi}}\right)e^{-\frac{\nu^2}{\xi}}}{\sum_{\nu=-\infty}^{\infty}e^{-\frac{\nu^2}{\xi}}} \tag{9.48}$$

(by (9.39), (9.44))

$$\leq \omega_1\left(f,\sqrt{\xi}\right)\left(\frac{\sqrt{\pi\xi}\left(1-\text{erf}\left(\frac{1}{\sqrt{\xi}}\right)\right)+e^{-\frac{1}{\xi}}\left(\sqrt{\xi}+2+\frac{2}{\sqrt{\xi}}\right)+1}{\sqrt{\pi\xi}\left(1-\text{erf}\left(\frac{1}{\sqrt{\xi}}\right)\right)+1}\right) \tag{9.49}$$

$$= \omega_1\left(f,\sqrt{\xi}\right)\left(1+\frac{e^{-\frac{1}{\xi}}\left(\sqrt{\xi}+2+\frac{2}{\sqrt{\xi}}\right)}{\sqrt{\pi\xi}\left(1-\text{erf}\left(\frac{1}{\sqrt{\xi}}\right)\right)+1}\right).$$

So we have proved that

$$\left|\left(W_{\xi}^{*}f\right)(x)-f(x)\right|\leq\omega_{1}\left(f,\sqrt{\xi}\right)\left(1+\frac{e^{-\frac{1}{\xi}}\left(\sqrt{\xi}+2+\frac{2}{\sqrt{\xi}}\right)}{\sqrt{\pi\xi}\left(1-\operatorname{erf}\left(\frac{1}{\sqrt{\xi}}\right)\right)+1}\right), \tag{9.50}$$

$\forall\, x\in\mathbb{R}$, any $0<\xi\leq 1$.

The constant in the last inequality converges to 1, as $\xi\to 0+$.

The proof of the theorem is completed. □

It follows the global smoothness preservation property of W_{ξ}^{*}.

Theorem 9.4. *It holds*

$$\omega_{1}\left(W_{\xi}^{*}f,\delta\right)\leq\omega_{1}(f,\delta),\ \ \forall\,\delta>0. \tag{9.51}$$

Inequality (9.51) is sharp, attained by $f(x)=id(x)=x$.

Proof. We see that

$$\left(W_{\xi}^{*}f\right)(x)-\left(W_{\xi}^{*}f\right)(y)=\frac{\sum_{\nu=-\infty}^{\infty}\left(f(x+\nu)-f(y+\nu)\right)e^{-\frac{\nu^{2}}{\xi}}}{\sum_{\nu=-\infty}^{\infty}e^{-\frac{\nu^{2}}{\xi}}},\ \forall\,x,y\in\mathbb{R}. \tag{9.52}$$

Hence

$$\left|\left(W_{\xi}^{*}f\right)(x)-\left(W_{\xi}^{*}f\right)(y)\right|\leq\frac{\sum_{\nu=-\infty}^{\infty}\left|f(x+\nu)-f(y+\nu)\right|e^{-\frac{\nu^{2}}{\xi}}}{\sum_{\nu=-\infty}^{\infty}e^{-\frac{\nu^{2}}{\xi}}}$$

$$\leq\omega_{1}(f,|x-y|)\left(\frac{\sum_{\nu=-\infty}^{\infty}e^{-\frac{\nu^{2}}{\xi}}}{\sum_{\nu=-\infty}^{\infty}e^{-\frac{\nu^{2}}{\xi}}}\right)=\omega_{1}(f,|x-y|),\ \forall\,x,y\in\mathbb{R}, \tag{9.53}$$

proving (9.51). Sharpness is obvious. □

Next we study the approximation properties of general discrete Poisson-Cauchy operators.

Theorem 9.5. *Let $f\in C_{U}(\mathbb{R})$, $0<\xi\leq 1$. Then*

$$\left\|M_{\xi}^{*}f-f\right\|\leq D(\xi)\,\omega_{1}(f,\xi), \tag{9.54}$$

where

$$D(\xi):=\left[1+4\xi^{2\alpha\beta}\left(\frac{\alpha\beta}{2\alpha\beta-1}\right)+\xi^{2\alpha\beta-1}\left(\frac{2\alpha\beta-1}{\alpha\beta-1}\right)\right]. \tag{9.55}$$

We have $\lim_{\xi\to 0+}D(\xi)=1$, and by $\lim_{\xi\to 0+}\omega_{1}(f,\xi)=0$, we get $M_{\xi}^{*}\to I$ uniformly with rates, as $\xi\to 0+$.

Proof. Here $0 < \xi \leq 1$, $\alpha \in \mathbb{N}$, $\beta > \frac{1}{\alpha}$, $x \in \mathbb{R}$. By [6], p. 397, formula 595, we have

$$\int_0^\infty \frac{1}{\left(t^{2\alpha} + \xi^{2\alpha}\right)^\beta} dt = \frac{\Gamma\left(\frac{1}{2\alpha}\right)\Gamma\left(\beta - \frac{1}{2\alpha}\right)}{2\Gamma(\beta)\,\alpha\xi^{2\alpha\beta-1}}. \tag{9.56}$$

Clearly $\left(\nu^{2\alpha} + \xi^{2\alpha}\right)^{-\beta}$ is decreasing, continuous and positive for $\nu \in [1, \infty)$. Hence by [5], we get

$$0 < \sum_{\nu=1}^{\infty} \left(\nu^{2\alpha} + \xi^{2\alpha}\right)^{-\beta} \leq \left(1 + \xi^{2\alpha}\right)^{-\beta} + \int_1^\infty \left(\nu^{2\alpha} + \xi^{2\alpha}\right)^{-\beta} d\nu \tag{9.57}$$

$$\leq \left(1 + \xi^{2\alpha}\right)^{-\beta} + \int_0^\infty \left(\nu^{2\alpha} + \xi^{2\alpha}\right)^{-\beta} d\nu = \left(1 + \xi^{2\alpha}\right)^{-\beta}$$

$$+ \frac{\Gamma\left(\frac{1}{2\alpha}\right)\Gamma\left(\beta - \frac{1}{2\alpha}\right)}{2\Gamma(\beta)\,\alpha\xi^{2\alpha\beta-1}} < \infty, \quad \forall\, \xi \in (0, 1].$$

Consequently we find convergence of

$$0 < S_1 := \sum_{\nu=-\infty}^{\infty} \left(\nu^{2\alpha} + \xi^{2\alpha}\right)^{-\beta} = \xi^{-2\alpha\beta} + 2\sum_{\nu=1}^{\infty} \left(\nu^{2\alpha} + \xi^{2\alpha}\right)^{-\beta}$$

$$\leq \xi^{-2\alpha\beta} + 2\left(1 + \xi^{2\alpha}\right)^{-\beta} + \frac{\Gamma\left(\frac{1}{2\alpha}\right)\Gamma\left(\beta - \frac{1}{2\alpha}\right)}{\Gamma(\beta)\,\alpha\xi^{2\alpha\beta-1}} < \infty, \quad \forall\, \xi \in (0, 1]. \tag{9.58}$$

Similarly we have

$$\sum_{\nu=1}^{\infty} \left(\nu^{2\alpha} + \xi^{2\alpha}\right)^{-\beta} \geq \int_1^\infty \left(\nu^{2\alpha} + \xi^{2\alpha}\right)^{-\beta} d\nu, \tag{9.59}$$

and

$$\sum_{\nu=-\infty}^{\infty} \left(\nu^{2\alpha} + \xi^{2\alpha}\right)^{-\beta} \geq \xi^{-2\alpha\beta} + 2\int_1^\infty \left(\nu^{2\alpha} + \xi^{2\alpha}\right)^{-\beta} d\nu. \tag{9.60}$$

That is

$$0 < \frac{1}{\sum_{\nu=-\infty}^{\infty} \left(\nu^{2\alpha} + \xi^{2\alpha}\right)^{-\beta}} \leq \frac{1}{\xi^{-2\alpha\beta} + 2\int_1^\infty \left(\nu^{2\alpha} + \xi^{2\alpha}\right)^{-\beta} d\nu} < \xi^{2\alpha\beta}. \tag{9.61}$$

That

$$0 < \frac{1}{S_1} < \xi^{2\alpha\beta} \to 0, \text{ as } \xi \to 0+. \tag{9.62}$$

Hence

$$\lim_{\xi \to 0+} \frac{1}{S_1} = 0. \tag{9.63}$$

Call $g(\nu) := \nu\left(\nu^{2\alpha} + \xi^{2\alpha}\right)^{-\beta}$, $\nu \in [1, \infty)$. We have that

$$g'(\nu) = \left(\nu^{2\alpha} + \xi^{2\alpha}\right)^{-\beta}\left[1 - \left(\frac{2\alpha\beta\nu^{2\alpha}}{\nu^{2\alpha} + \xi^{2\alpha}}\right)\right] \leq 0, \tag{9.64}$$

iff $1-\left(\frac{2\alpha\beta\nu^{2\alpha}}{\nu^{2\alpha}+\xi^{2\alpha}}\right)\leq 0$, iff $\nu^{2\alpha}+\xi^{2\alpha}\leq 2\alpha\beta\nu^{2\alpha}$, iff $\xi^{2\alpha}\leq\nu^{2\alpha}(2\alpha\beta-1)$, which is true because $2\alpha\beta-1\geq 1$ and $\nu^{2\alpha}(2\alpha\beta-1)\geq 1\geq\xi^{2\alpha}$. That is g is decreasing, positive and continuous on $[1,\infty)$.

Hence $\left(1+\frac{\nu}{\xi}\right)\left(\nu^{2\alpha}+\xi^{2\alpha}\right)^{-\beta}$ is decreasing, positive and continuous on $[1,\infty)$.

Thus again by [5] we derive

$$\sum_{\nu=1}^{\infty}\left(1+\frac{\nu}{\xi}\right)\left(\nu^{2\alpha}+\xi^{2\alpha}\right)^{-\beta} \tag{9.65}$$

$$\leq\left(1+\frac{1}{\xi}\right)\left(1+\xi^{2\alpha}\right)^{-\beta}+\int_{1}^{\infty}\left(1+\frac{\nu}{\xi}\right)\left(\nu^{2\alpha}+\xi^{2\alpha}\right)^{-\beta}d\nu.$$

We further notice that

$$\int_{1}^{\infty}\left(1+\frac{\nu}{\xi}\right)\left(\nu^{2\alpha}+\xi^{2\alpha}\right)^{-\beta}d\nu$$

$$=\int_{1}^{\infty}\left(\nu^{2\alpha}+\xi^{2\alpha}\right)^{-\beta}d\nu+\frac{1}{\xi}\int_{1}^{\infty}\nu\left(\nu^{2\alpha}+\xi^{2\alpha}\right)^{-\beta}d\nu$$

$$<\int_{1}^{\infty}\nu^{-2\alpha\beta}d\nu+\frac{1}{\xi}\int_{1}^{\infty}\nu^{-2\alpha\beta+1}d\nu=\left(\frac{1}{2\alpha\beta-1}\right) \tag{9.66}$$

$$+\left(\frac{1}{2\xi\left(\alpha\beta-1\right)}\right)<\infty,\ \forall\ \xi\in(0,1].$$

So that

$$\sum_{\nu=1}^{\infty}\left(1+\frac{\nu}{\xi}\right)\left(\nu^{2\alpha}+\xi^{2\alpha}\right)^{-\beta} \tag{9.67}$$

$$<\left(1+\frac{1}{\xi}\right)\left(1+\xi^{2\alpha}\right)^{-\beta}+\left(\frac{1}{2\alpha\beta-1}\right)+\left(\frac{1}{2\xi\left(\alpha\beta-1\right)}\right)<\infty,\ \forall\ \xi\in(0,1].$$

Consequently we obtain

$$0<S_2:=\sum_{\nu=-\infty}^{\infty}\left(1+\frac{|\nu|}{\xi}\right)\left(\nu^{2\alpha}+\xi^{2\alpha}\right)^{-\beta} \tag{9.68}$$

$$=\xi^{-2\alpha\beta}+2\sum_{\nu=1}^{\infty}\left(1+\frac{\nu}{\xi}\right)\left(\nu^{2\alpha}+\xi^{2\alpha}\right)^{-\beta}$$

$$<\xi^{-2\alpha\beta}+2\left(1+\frac{1}{\xi}\right)\left(1+\xi^{2\alpha}\right)^{-\beta}+\left(\frac{2}{2\alpha\beta-1}\right)+\left(\frac{1}{\xi\left(\alpha\beta-1\right)}\right)$$

$$<\frac{1}{\xi^{2\alpha\beta}}+2\left(1+\frac{1}{\xi}\right)+\left(\frac{2}{2\alpha\beta-1}\right)+\left(\frac{1}{\xi\left(\alpha\beta-1\right)}\right)=:\varphi\left(\xi\right).$$

So that

$$0 < S_2 < \varphi(\xi) < \infty, \; \forall\, \xi \in (0,1], \tag{9.69}$$

and

$$0 < \frac{S_2}{S_1} \overset{(9.62)}{<} \xi^{2\alpha\beta}\varphi(\xi)$$

$$= 1 + 2\xi^{2\alpha\beta}\left(1 + \frac{1}{2\alpha\beta - 1}\right) + \xi^{2\alpha\beta-1}\left(2 + \frac{1}{\alpha\beta - 1}\right)$$

$$= \left[1 + 4\xi^{2\alpha\beta}\left(\frac{\alpha\beta}{2\alpha\beta - 1}\right) + \xi^{2\alpha\beta-1}\left(\frac{2\alpha\beta - 1}{\alpha\beta - 1}\right)\right] \to 1, \text{ as } \xi \to 0+. \tag{9.70}$$

Hence

$$0 < \lim_{\xi \to 0+} \frac{S_2}{S_1} < 1. \tag{9.71}$$

Finally we have that

$$M_\xi^*(f,x) - f(x) = \frac{\sum_{\nu=-\infty}^{\infty} \left(f(x+\nu) - f(x)\right)\left(\nu^{2\alpha} + \xi^{2\alpha}\right)^{-\beta}}{\sum_{\nu=-\infty}^{\infty}\left(\nu^{2\alpha} + \xi^{2\alpha}\right)^{-\beta}}, \tag{9.72}$$

and

$$\left|M_\xi^*(f,x) - f(x)\right| \le \frac{\sum_{\nu=-\infty}^{\infty} \left|f(x+\nu) - f(x)\right|\left(\nu^{2\alpha} + \xi^{2\alpha}\right)^{-\beta}}{\sum_{\nu=-\infty}^{\infty}\left(\nu^{2\alpha} + \xi^{2\alpha}\right)^{-\beta}} \tag{9.73}$$

$$\le \frac{\sum_{\nu=-\infty}^{\infty} \omega_1\left(f, \xi\frac{|\nu|}{\xi}\right)\left(\nu^{2\alpha} + \xi^{2\alpha}\right)^{-\beta}}{\sum_{\nu=-\infty}^{\infty}\left(\nu^{2\alpha} + \xi^{2\alpha}\right)^{-\beta}}$$

$$\le \omega_1(f,\xi)\left(\frac{\sum_{\nu=-\infty}^{\infty}\left(1 + \frac{|\nu|}{\xi}\right)\left(\nu^{2\alpha} + \xi^{2\alpha}\right)^{-\beta}}{\sum_{\nu=-\infty}^{\infty}\left(\nu^{2\alpha} + \xi^{2\alpha}\right)^{-\beta}}\right) = \left(\frac{S_2}{S_1}\right)\omega_1(f,\xi) \tag{9.74}$$

$$\overset{(9.70)}{\le} \left[1 + 4\xi^{2\alpha\beta}\left(\frac{\alpha\beta}{2\alpha\beta - 1}\right) + \xi^{2\alpha\beta-1}\left(\frac{2\alpha\beta - 1}{\alpha\beta - 1}\right)\right]\omega_1(f,\xi).$$

We have derived

$$\left|M_\xi^*(f,x) - f(x)\right| \le \left[1 + 4\xi^{2\alpha\beta}\left(\frac{\alpha\beta}{2\alpha\beta - 1}\right) + \xi^{2\alpha\beta-1}\left(\frac{2\alpha\beta - 1}{\alpha\beta - 1}\right)\right]\omega_1(f,\xi), \tag{9.75}$$

$\forall\, x \in \mathbb{R}$, $\forall\, \xi \in (0,1]$, proving the claim. □

It follows the global smoothness preservation property of M_ξ^*.

Theorem 9.6. *It holds*

$$\omega_1\left(M_\xi^* f, \delta\right) \le \omega_1(f,\delta), \;\; \forall\, \delta > 0. \tag{9.76}$$

Inequality (9.76) is sharp, attained by $f(x) = id(x) = x$.

Proof. Similar to the proof of Theorem 9.4. □

We continue with

Theorem 9.7. *It holds*

$$\left\|\theta_{\xi}^{*}f-f\right\|_{\infty}\leq\left(\frac{2}{3}\pi^{2}K^{2\beta}\right)\xi^{2\beta-1}\omega_{1}\left(f,\xi\right)\rightarrow0,\ as\ \xi\rightarrow0+. \tag{9.77}$$

Proof. Here we use a lot

$$\sum_{\nu=1}^{\infty}\frac{1}{\nu^{2}}=\frac{\pi^{2}}{6}\quad\text{(Euler, 1741)}. \tag{9.78}$$

We have

$$\theta_{\xi}^{*}\left(f,x\right)-f\left(x\right)=\frac{\xi^{2\beta}\sum_{\nu\in\mathbb{Z}-\{0\}}\left(f\left(x+\nu\right)-f\left(x\right)\right)\left(\frac{\varphi\left(\frac{\nu}{\xi}\right)}{\nu}\right)^{2\beta}}{1+\xi^{2\beta}\sum_{\nu\in\mathbb{Z}-\{0\}}\left(\frac{\varphi\left(\frac{\nu}{\xi}\right)}{\nu}\right)^{2\beta}}. \tag{9.79}$$

Hence

$$\left|\theta_{\xi}^{*}\left(f,x\right)-f\left(x\right)\right|\leq\frac{\xi^{2\beta}\sum_{\nu\in\mathbb{Z}-\{0\}}\left|f\left(x+\nu\right)-f\left(x\right)\right|\left(\frac{\varphi\left(\frac{\nu}{\xi}\right)}{\nu}\right)^{2\beta}}{1+\xi^{2\beta}\sum_{\nu\in\mathbb{Z}-\{0\}}\left(\frac{\varphi\left(\frac{\nu}{\xi}\right)}{\nu}\right)^{2\beta}}$$

$$\leq\frac{\xi^{2\beta}\sum_{\nu\in\mathbb{Z}-\{0\}}\omega_{1}\left(f,\xi\frac{|\nu|}{\xi}\right)\left(\frac{\varphi\left(\frac{\nu}{\xi}\right)}{\nu}\right)^{2\beta}}{1+\xi^{2\beta}\sum_{\nu\in\mathbb{Z}-\{0\}}\left(\frac{\varphi\left(\frac{\nu}{\xi}\right)}{\nu}\right)^{2\beta}} \tag{9.80}$$

$$\leq\frac{\xi^{2\beta}\omega_{1}\left(f,\xi\right)\sum_{\nu\in\mathbb{Z}-\{0\}}\left(1+\frac{|\nu|}{\xi}\right)\left(\frac{\varphi\left(\frac{\nu}{\xi}\right)}{\nu}\right)^{2\beta}}{1+\xi^{2\beta}\sum_{\nu\in\mathbb{Z}-\{0\}}\left(\frac{\varphi\left(\frac{\nu}{\xi}\right)}{\nu}\right)^{2\beta}} \tag{9.81}$$

$$\leq\frac{\xi^{2\beta}\omega_{1}\left(f,\xi\right)K^{2\beta}\left(\sum_{\nu\in\mathbb{Z}-\{0\}}\left(1+\frac{|\nu|}{\xi}\right)\frac{1}{\nu^{2\beta}}\right)}{1+\xi^{2\beta}\sum_{\nu\in\mathbb{Z}-\{0\}}\left(\frac{\varphi\left(\frac{\nu}{\xi}\right)}{\nu}\right)^{2\beta}}=:(*).$$

We observe that

$$0\leq S_{2}:=\sum_{\nu\in\mathbb{Z}-\{0\}}\left(\frac{\varphi\left(\frac{\nu}{\xi}\right)}{\nu}\right)^{2\beta}\leq K^{2\beta}\sum_{\nu\in\mathbb{Z}-\{0\}}\frac{1}{\nu^{2\beta}}$$

$$= 2K^{2\beta} \sum_{\nu=1}^{\infty} \frac{1}{\nu^{2\beta}} < 2K^{2\beta} \sum_{\nu=1}^{\infty} \frac{1}{\nu^{2}} = \frac{K^{2\beta}\pi^2}{3}. \tag{9.82}$$

Thus

$$0 \leq S_2 \leq \frac{K^{2\beta}\pi^2}{3}. \tag{9.83}$$

So that

$$1 \leq 1 + \xi^{2\beta} S_2 \leq 1 + \frac{\xi^{2\beta} K^{2\beta}\pi^2}{3} < \infty, \quad \forall\, \xi > 0. \tag{9.84}$$

That is

$$0 < \frac{1}{1 + \xi^{2\beta} S_2} \leq 1, \tag{9.85}$$

with

$$\lim_{\xi \to 0+} \frac{1}{1 + \xi^{2\beta} S_2} = 1. \tag{9.86}$$

Consequently it holds

$$(*) \leq 2\xi^{2\beta} \omega_1(f, \xi) K^{2\beta} \left(\sum_{\nu=1}^{\infty} \left(1 + \frac{\nu}{\xi}\right) \frac{1}{\nu^{2\beta}} \right)$$

$$= 2\xi^{2\beta} \omega_1(f, \xi) K^{2\beta} \left(\sum_{\nu=1}^{\infty} \frac{1}{\nu^{2\beta}} + \frac{1}{\xi} \sum_{\nu=1}^{\infty} \frac{1}{\nu^{2\beta-1}} \right)$$

$$\leq 2\xi^{2\beta} \omega_1(f, \xi) K^{2\beta} \left(\sum_{\nu=1}^{\infty} \frac{1}{\nu^{2}} + \frac{1}{\xi} \sum_{\nu=1}^{\infty} \frac{1}{\nu^{2}} \right)$$

$$= \frac{\xi^{2\beta}\pi^2}{3} \omega_1(f, \xi) K^{2\beta} \left(1 + \frac{1}{\xi}\right) \leq \frac{2\xi^{2\beta-1}\pi^2}{3} K^{2\beta} \omega_1(f, \xi). \tag{9.87}$$

We have proved that

$$\left|\theta_{\xi}^{*}(f, x) - f(x)\right| \leq \left(\frac{2}{3}\pi^2 K^{2\beta}\right) \xi^{2\beta-1} \omega_1(f, \xi) \to 0, \text{ as } \xi \to 0+. \tag{9.88}$$

The proof is completed. □

Example 9.1. In Theorem 9.7 we can take φ to be sine, cosine with $K = 1$.

Theorem 9.8. *It holds*

$$\omega_1\left(\theta_{\xi}^{*} f, \delta\right) \leq \omega_1(f, \delta), \quad \forall\, \delta > 0. \tag{9.89}$$

Inequality (9.89) is attained by $f = id$.

We finish by studying T_{ξ}^{*}.

Theorem 9.9. *It holds*

$$\left\|T_{\xi}^{*}f-f\right\|_{\infty}\leq\left(\frac{2\pi^{2}A}{3}\right)\xi^{2\beta-1}\omega_{1}\left(f,\xi\right)\rightarrow 0,\ as\ \xi\rightarrow 0+. \tag{9.90}$$

Theorem 9.10. *It holds*

$$\omega_{1}\left(T_{\xi}^{*}f,\delta\right)\leq\omega_{1}\left(f,\delta\right),\quad\forall\,\delta>0. \tag{9.91}$$

Inequality (9.91) is attained by $f=id$.

Proof of Theorem 9.9. We have

$$T_{\xi}^{*}\left(f,x\right)-f\left(x\right)=\frac{\sum_{\nu\in\mathbb{Z}-\{0\}}\left(f\left(x+\nu\right)-f\left(x\right)\right)\left(\frac{\varphi\left(\frac{\nu}{\xi}\right)}{\left(\frac{\nu}{\xi}\right)^{4\beta}}\right)}{1+\sum_{\nu\in\mathbb{Z}-\{0\}}\left(\frac{\varphi\left(\frac{\nu}{\xi}\right)}{\left(\frac{\nu}{\xi}\right)^{4\beta}}\right)}. \tag{9.92}$$

Thus

$$\left|T_{\xi}^{*}\left(f,x\right)-f\left(x\right)\right|\leq\frac{\sum_{\nu\in\mathbb{Z}-\{0\}}\left|f\left(x+\nu\right)-f\left(x\right)\right|\left(\frac{\varphi\left(\frac{\nu}{\xi}\right)}{\left(\frac{\nu}{\xi}\right)^{4\beta}}\right)}{1+\sum_{\nu\in\mathbb{Z}-\{0\}}\left(\frac{\varphi\left(\frac{\nu}{\xi}\right)}{\left(\frac{\nu}{\xi}\right)^{4\beta}}\right)}$$

$$\leq\frac{\sum_{\nu\in\mathbb{Z}-\{0\}}\omega_{1}\left(f,\xi\frac{|\nu|}{\xi}\right)\left(\frac{\varphi\left(\frac{\nu}{\xi}\right)}{\left(\frac{\nu}{\xi}\right)^{4\beta}}\right)}{1+\sum_{\nu\in\mathbb{Z}-\{0\}}\left(\frac{\varphi\left(\frac{\nu}{\xi}\right)}{\left(\frac{\nu}{\xi}\right)^{4\beta}}\right)} \tag{9.93}$$

$$\leq\frac{\omega_{1}\left(f,\xi\right)\sum_{\nu\in\mathbb{Z}-\{0\}}\left(1+\frac{|\nu|}{\xi}\right)\left(\frac{\varphi\left(\frac{\nu}{\xi}\right)}{\left(\frac{\nu}{\xi}\right)^{4\beta}}\right)}{1+\sum_{\nu\in\mathbb{Z}-\{0\}}\left(\frac{\varphi\left(\frac{\nu}{\xi}\right)}{\left(\frac{\nu}{\xi}\right)^{4\beta}}\right)} \tag{9.94}$$

$$\leq\frac{\omega_{1}\left(f,\xi\right)\sum_{\nu\in\mathbb{Z}-\{0\}}\left(1+\frac{|\nu|}{\xi}\right)\frac{A}{\left(\frac{\nu}{\xi}\right)^{2\beta}}}{1+\sum_{\nu\in\mathbb{Z}-\{0\}}\left(\frac{\varphi\left(\frac{\nu}{\xi}\right)}{\left(\frac{\nu}{\xi}\right)^{4\beta}}\right)}$$

$$=\frac{A\omega_{1}\left(f,\xi\right)\xi^{2\beta}\sum_{\nu\in\mathbb{Z}-\{0\}}\left(1+\frac{|\nu|}{\xi}\right)\frac{1}{\nu^{2\beta}}}{1+\sum_{\nu\in\mathbb{Z}-\{0\}}\left(\frac{\varphi\left(\frac{\nu}{\xi}\right)}{\left(\frac{\nu}{\xi}\right)^{4\beta}}\right)} \tag{9.95}$$

$$=\frac{2A\omega_{1}\left(f,\xi\right)\xi^{2\beta}\sum_{\nu=1}^{\infty}\left(1+\frac{\nu}{\xi}\right)\frac{1}{\nu^{2\beta}}}{1+\sum_{\nu\in\mathbb{Z}-\{0\}}\left(\frac{\varphi\left(\frac{\nu}{\xi}\right)}{\left(\frac{\nu}{\xi}\right)^{4\beta}}\right)}$$

$$= \frac{2A\omega_1(f,\xi)\,\xi^{2\beta}\left[\sum_{\nu=1}^{\infty}\frac{1}{\nu^{2\beta}}+\frac{1}{\xi}\sum_{\nu=1}^{\infty}\frac{1}{\nu^{2\beta-1}}\right]}{1+\sum_{\nu\in\mathbb{Z}-\{0\}}\left(\frac{\varphi\left(\frac{\nu}{\xi}\right)}{\left(\frac{\nu}{\xi}\right)^{4\beta}}\right)} \tag{9.96}$$

$$\leq \frac{2A\omega_1(f,\xi)\,\xi^{2\beta}\left[\frac{\pi^2}{6}+\frac{1}{\xi}\frac{\pi^2}{6}\right]}{\left(1+\sum_{\nu\in\mathbb{Z}-\{0\}}\left(\frac{\varphi\left(\frac{\nu}{\xi}\right)}{\left(\frac{\nu}{\xi}\right)^{4\beta}}\right)\right)} \leq \frac{\left(\frac{2\pi^2}{3}A\right)\omega_1(f,\xi)\,\xi^{2\beta-1}}{1+\sum_{\nu\in\mathbb{Z}-\{0\}}\frac{\varphi\left(\frac{\nu}{\xi}\right)}{\left(\frac{\nu}{\xi}\right)^{4\beta}}} =: (*). \tag{9.97}$$

We see that

$$1+\sum_{\nu\in\mathbb{Z}-\{0\}}\frac{\varphi\left(\frac{\nu}{\xi}\right)}{\left(\frac{\nu}{\xi}\right)^{4\beta}} \leq 1+A\sum_{\nu\in\mathbb{Z}-\{0\}}\frac{1}{\left(\frac{\nu}{\xi}\right)^{2\beta}} \tag{9.98}$$

$$= 1+A\xi^{2\beta}\sum_{\nu\in\mathbb{Z}-\{0\}}\frac{1}{\nu^{2\beta}} = 1+2A\xi^{2\beta}\sum_{\nu=1}^{\infty}\frac{1}{\nu^{2\beta}}$$

$$< 1+2A\xi^{2\beta}\left(\sum_{\nu=1}^{\infty}\frac{1}{\nu^2}\right) = 1+\frac{A\xi^{2\beta}\pi^2}{3} < \infty, \quad \forall\, \xi > 0. \tag{9.99}$$

That is

$$1 \leq 1+\sum_{\nu\in\mathbb{Z}-\{0\}}\frac{\varphi\left(\frac{\nu}{\xi}\right)}{\left(\frac{\nu}{\xi}\right)^{4\beta}} < 1+\frac{A\pi^2\xi^{2\beta}}{3} < \infty, \quad \forall\, \xi > 0. \tag{9.100}$$

Also $1+\frac{A\pi^2\xi^{2\beta}}{3} \to 1$, as $\xi \to 0+$, that is

$$\lim_{\xi\to 0+}\left(1+\sum_{\nu\in\mathbb{Z}-\{0\}}\frac{\varphi\left(\frac{\nu}{\xi}\right)}{\left(\frac{\nu}{\xi}\right)^{4\beta}}\right) = 1. \tag{9.101}$$

Furthermore we have

$$0 < \frac{1}{1+\sum_{\nu\in\mathbb{Z}-\{0\}}\frac{\varphi\left(\frac{\nu}{\xi}\right)}{\left(\frac{\nu}{\xi}\right)^{4\beta}}} \leq 1. \tag{9.102}$$

Hence

$$(*) \leq \left(\frac{2\pi^2}{3}A\right)\omega_1(f,\xi)\,\xi^{2\beta-1}. \tag{9.103}$$

We have proved

$$\left|T_\xi^*(f,x)-f(x)\right| \leq \left(\frac{2\pi^2 A}{3}\right)\xi^{2\beta-1}\omega_1(f,\xi) \to 0, \text{ as } \xi\to 0+, \ \forall\, x\in\mathbb{R}, \tag{9.104}$$

proving the claim. □

Note. All estimates of this chapter are also true for $f \in C_b(\mathbb{R})$, continuous and bounded functions on $\mathbb{R}$. However the convergences fail if f is not uniformly continuous.

Bibliography

1. G. Anastassiou and R. Mezei, *Approximation by Singular Integrals*, Cambridge Scientific Publishers, Cambridge, U.K., 2013.
2. G. A. Anastassiou, *Approximation by discrete singular operators*, CUBO, Vol. 15, No. 1 (2013), 97-112.
3. R.A. DeVore and G.G. Lorentz, *Constructive Approximation*, Springer-Verlag, New York, Heidelberg, 1993.
4. I. Favard, *Sur les multiplicateurs d'interpolation*, J. Math. Pures Appl., IX 23 (1944), 219-247.
5. F. Smarandache, *A triple inequality with series and improper integrals*, arxiv.org/ftp/math/papers/0605/0605027.pdf, 2006.
6. D. Zwillinger, *CRC Standard Mathematical Tables and Formulae*, 30th edition, Chapman & Hall/CRC, Boca Raton, 1995.

Chapter 10

On Discrete Approximation by Gauss-Weierstrass and Picard Type Operators

Here we study basic approximation properties with rates of our discrete versions of Gauss-Weierstrass and Picard singular operators. We prove quantitatively pointwise and uniform convergence of these operators to the unit operator. Also these operators fulfill the global smoothness preservation property. The discussed operators act on continuous functions over the real line. It follows [4].

10.1 Preliminaries

This chapter is motivated mainly by [8], where J. Favard in 1944 introduced the discrete version of Gauss-Weierstrass operator

$$(F_n f)(x) = \frac{1}{\sqrt{\pi n}} \sum_{\nu=-\infty}^{\infty} f\left(\frac{\nu}{n}\right) \exp\left(-n\left(\frac{\nu}{n} - x\right)^2\right), \tag{10.1}$$

$n \in \mathbb{N}$, which has the property that $(F_n f)(x)$ converges to $f(x)$ pointwise for each $x \in \mathbb{R}$, and uniformly on any compact subinterval of $\mathbb{R}$, for each continuous function f ($f \in C(\mathbb{R})$) that fulfills $|f(t)| \leq Ae^{Bt^2}$, $t \in \mathbb{R}$, where A, B are positive constants.

The well-known Gauss-Weierstrass singular convolution integral operator is

$$(W_n f)(x) = \sqrt{\frac{n}{\pi}} \int_{-\infty}^{\infty} f(u) \exp\left(-n(u-x)^2\right) du. \tag{10.2}$$

In [1], [2], the authors studied extensively the approximation properties of generalizations of (10.1) and in particular they established important asymptotic expansions.

We are also motivated by [3] and [6].

In [5] the authors studied in depth and length the approximation properties by (10.2), as well as of Picard and Poisson-Cauchy integral singular operators.

Let $f : \mathbb{R} \to \mathbb{R}$ be a function which is uniformly continuous ($f \in C_U(\mathbb{R})$). Following [7], p. 40-41, we define the first modulus of continuity,

$$\omega_1(f,t) := \sup_{\substack{x,y \in \mathbb{R} \\ |x-y| \leq t}} |f(x) - f(y)|, \quad t \geq 0. \tag{10.3}$$

The function ω_1 is continuous at $t = 0$ if and only if f is uniformly continuous on $\mathbb{R}$. So when $f \in C_U(\mathbb{R})$ we get $\omega_1(f,t) \to \omega_1(f,0) = 0$, as $t \to 0$. It also holds

$$\omega_1(f, \lambda t) \leq (\lambda + 1)\,\omega_1(f,t), \quad \lambda \geq 0. \tag{10.4}$$

Clearly $\omega_1(f,t)$ is finite for each $t \geq 0$.

In the case of $f : \mathbb{R} \to \mathbb{R}$ continuous and bounded ($f \in C_b(\mathbb{R})$), $\omega_1(f,t)$ is also finite for each $t \geq 0$ and (10.4) is still valid, but we do not have necessarily continuity at zero of $\omega_1(f,\cdot)$.

In this chapter we study the pointwise and uniform convergence with rates, to the unit operator I, of the following operators W_ξ and P_ξ, that are our discrete versions of Gauss-Weierstrass and Picard integral convolution operators, respectively.

We study also the global smoothness preservation property shared by W_ξ, P_ξ.

So let $0 < \xi \leq 1$, $\xi \to 0+$, $f \in C_U(\mathbb{R})$ or $f \in C_b(\mathbb{R})$.

We define

$$(W_\xi f)(x) := \frac{\sum_{\nu=-\infty}^{\infty} f(x+\nu)\, e^{-\frac{\nu^2}{\xi}}}{\sqrt{\pi\xi}\left(1 - \operatorname{erf}\left(\frac{1}{\sqrt{\xi}}\right)\right) + 1}, \tag{10.5}$$

where $\operatorname{erf}(x) = \frac{2}{\sqrt{\pi}} \int_0^x e^{-t^2}dt$, $\operatorname{erf}(\infty) = 1$.

We also define

$$(P_\xi f)(x) = \frac{\sum_{\nu=-\infty}^{\infty} f(x+\nu)\, e^{-\frac{|\nu|}{\xi}}}{1 + 2\xi e^{-\frac{1}{\xi}}}, \tag{10.6}$$

which is the discrete analog of well-known Picard integral

$$(\Pi_\xi f)(x) = \frac{1}{2\xi} \int_{-\infty}^{\infty} f(x+t)\, e^{-\frac{|t|}{\xi}}\, dt. \tag{10.7}$$

10.2 Main Results

First we study the approximation properties of discrete Gauss-Weierstrass operators W_ξ.

Theorem 10.1. *Let $f \in C_U(\mathbb{R})$ or $f \in C_b(\mathbb{R})$, $x \in \mathbb{R}$, $0 < \xi \leq 1$. Then*

i)

$$|(W_\xi f)(x) - f(x)| \leq \frac{1}{\left(\sqrt{\pi\xi}\left(1 - \operatorname{erf}\left(\frac{1}{\sqrt{\xi}}\right)\right) + 1\right)}$$

$$\cdot \Bigg[2\,|f(x)|\, e^{-\frac{1}{\xi}} + \Bigg\{ \sqrt{\pi\xi}\left(1 - \operatorname{erf}\left(\frac{1}{\sqrt{\xi}}\right)\right)$$

$$+ e^{-\frac{1}{\xi}}\left(\sqrt{\xi} + 2 + \frac{2}{\sqrt{\xi}}\right) + 1 \Bigg\}\, \omega_1\left(f, \sqrt{\xi}\right) \Bigg], \tag{10.8}$$

and for $f \in C_U(\mathbb{R})$, as $\xi \to 0+$, we obtain $(W_\xi f)(x) \to f(x)$, pointwise,

ii) for $f \in C_b(\mathbb{R})$ we derive

$$\|W_\xi f - f\|_\infty \le \frac{1}{\sqrt{\pi\xi}\left(1 - \operatorname{erf}\left(\frac{1}{\sqrt{\xi}}\right)\right) + 1}$$

$$\cdot \left[2\|f\|_\infty e^{-\frac{1}{\xi}} + \left\{\sqrt{\pi\xi}\left(1 - \operatorname{erf}\left(\frac{1}{\sqrt{\xi}}\right)\right)\right.\right.$$

$$\left.\left. + e^{-\frac{1}{\xi}}\left(\sqrt{\xi} + 2 + \frac{2}{\sqrt{\xi}}\right) + 1\right\}\omega_1\left(f, \sqrt{\xi}\right)\right], \tag{10.9}$$

and for $f \in C_U(\mathbb{R}) \cap C_b(\mathbb{R})$, as $\xi \to 0+$, we get $W_\xi f \to f$ uniformly, i.e. $W_\xi \xrightarrow{u} I$.

Proof. Since $e^{-\frac{\nu^2}{\xi}}$ is positive, continuous and decreasing we get

$$\sum_{\nu=1}^{\infty} e^{-\frac{\nu^2}{\xi}} - e^{-\frac{1}{\xi}} \le \int_1^\infty e^{-\frac{\nu^2}{\xi}} d\nu \le \sum_{\nu=1}^{\infty} e^{-\frac{\nu^2}{\xi}}, \tag{10.10}$$

that is

$$0 \le \sum_{\nu=1}^{\infty} e^{-\frac{\nu^2}{\xi}} - \int_1^\infty e^{-\frac{\nu^2}{\xi}} d\nu \le e^{-\frac{1}{\xi}}, \tag{10.11}$$

and

$$\left(\sum_{\nu=1}^{\infty} e^{-\frac{\nu^2}{\xi}} - \int_1^\infty e^{-\frac{\nu^2}{\xi}} d\nu\right) \to 0, \text{ as } \xi \to 0+.$$

We easily obtain that

$$2\int_1^\infty e^{-\frac{\nu^2}{\xi}} d\nu + 1 = \sqrt{\pi\xi}\left(1 - \operatorname{erf}\left(\frac{1}{\sqrt{\xi}}\right)\right) + 1 \to 1, \text{ as } \xi \to 0+. \tag{10.12}$$

Also we have

$$\sum_{\nu=-\infty}^{\infty} e^{-\frac{\nu^2}{\xi}} = 1 + 2\sum_{\nu=1}^{\infty} e^{-\frac{\nu^2}{\xi}} < \infty.$$

Similarly, we obtain that

$$\sum_{\nu=1}^{\infty}\left(1 + \frac{\nu}{\sqrt{\xi}}\right) e^{-\frac{\nu^2}{\xi}} \le \int_1^\infty \left(1 + \frac{\nu}{\sqrt{\xi}}\right) e^{-\frac{\nu^2}{\xi}} d\nu + \left(1 + \frac{1}{\sqrt{\xi}}\right) e^{-\frac{1}{\xi}}$$

$$= \frac{\sqrt{\pi\xi}}{2}\left(1 - \operatorname{erf}\left(\frac{1}{\sqrt{\xi}}\right)\right) + e^{-\frac{1}{\xi}}\left(\frac{\sqrt{\xi}}{2} + 1 + \frac{1}{\sqrt{\xi}}\right) \to 0, \text{ as } \xi \to 0+. \tag{10.13}$$

Consequently it holds

$$\sum_{\nu=-\infty}^{\infty}\left(1 + \frac{|\nu|}{\sqrt{\xi}}\right) e^{-\frac{\nu^2}{\xi}} \le \sqrt{\pi\xi}\left(1 - \operatorname{erf}\left(\frac{1}{\sqrt{\xi}}\right)\right)$$

$$+e^{-\frac{1}{\xi}}\left(\sqrt{\xi}+2+\frac{2}{\sqrt{\xi}}\right)+1\to 1,\text{ as }\xi\to 0+. \tag{10.14}$$

We notice that

$$(W_\xi f)(x)=\frac{\sum_{\nu=-\infty}^{\infty} f(x+\nu)\,e^{-\frac{\nu^2}{\xi}}}{1+2\int_1^\infty e^{-\frac{\nu^2}{\xi}}d\nu}, \tag{10.15}$$

and

$$(W_\xi f)(x)-f(x) \tag{10.16}$$

$$=\frac{2f(x)\left[\sum_{\nu=1}^{\infty} e^{-\frac{\nu^2}{\xi}}-\int_1^\infty e^{-\frac{\nu^2}{\xi}}d\nu\right]+\sum_{\nu=-\infty}^{\infty}\left(f(x+\nu)-f(x)\right)e^{-\frac{\nu^2}{\xi}}}{1+2\int_1^\infty e^{-\frac{\nu^2}{\xi}}d\nu}.$$

Therefore we get

$$|(W_\xi f)(x)-f(x)|$$

$$\overset{((10.11),(10.16))}{\leq}\frac{2|f(x)|\,e^{-\frac{1}{\xi}}+\sum_{\nu=-\infty}^{\infty}|f(x+\nu)-f(x)|\,e^{-\frac{\nu^2}{\xi}}}{1+2\int_1^\infty e^{-\frac{\nu^2}{\xi}}d\nu} \tag{10.17}$$

$$\leq\frac{\left(2|f(x)|\,e^{-\frac{1}{\xi}}+\sum_{\nu=-\infty}^{\infty}\omega_1(f,|\nu|)\,e^{-\frac{\nu^2}{\xi}}\right)}{\left(\sqrt{\pi\xi}\left(1-\operatorname{erf}\left(\frac{1}{\sqrt{\xi}}\right)\right)+1\right)}$$

$$\overset{(10.4)}{\leq}\frac{\left(2|f(x)|\,e^{-\frac{1}{\xi}}+\omega_1(f,\sqrt{\xi})\sum_{\nu=-\infty}^{\infty}\left(1+\frac{|\nu|}{\sqrt{\xi}}\right)e^{-\frac{\nu^2}{\xi}}\right)}{\left(\sqrt{\pi\xi}\left(1-\operatorname{erf}\left(\frac{1}{\sqrt{\xi}}\right)\right)+1\right)} \tag{10.18}$$

$$\overset{(10.14)}{\leq}\frac{\left[2|f(x)|\,e^{-\frac{1}{\xi}}+\omega_1(f,\sqrt{\xi})\left\{\sqrt{\pi\xi}\left(1-\operatorname{erf}\left(\frac{1}{\sqrt{\xi}}\right)\right)+e^{-\frac{1}{\xi}}\left(\sqrt{\xi}+2+\frac{2}{\sqrt{\xi}}\right)+1\right\}\right]}{\left(\sqrt{\pi\xi}\left(1-\operatorname{erf}\left(\frac{1}{\sqrt{\xi}}\right)\right)+1\right)}. \tag{10.19}$$

The proof is completed. □

Next we show the global smoothness preservation property of operators W_ξ. So that the approximation of $W_\xi f$ to f is nice, smooth and fit.

Theorem 10.2. *Let $f\in C(\mathbb{R})$ with $\omega_1(f,\delta)<\infty$, $\forall\,\delta>0$. Then*

$$\omega_1(W_\xi f,\delta)\leq\left[1+\frac{2}{e^{\frac{1}{\xi}}\left(\sqrt{\pi\xi}\left(1-\operatorname{erf}\left(\frac{1}{\sqrt{\xi}}\right)\right)+1\right)}\right]\omega_1(f,\delta), \tag{10.20}$$

$\forall\,\delta>0$, *any* $0<\xi\leq 1$.

Proof. Notice that

$$\sum_{\nu=-\infty}^{\infty} e^{-\frac{\nu^2}{\xi}} = 1 + 2\sum_{\nu=1}^{\infty} e^{-\frac{\nu^2}{\xi}} \tag{10.21}$$

$$\overset{(10.10)}{\leq} 1 + 2\int_1^{\infty} e^{-\frac{\nu^2}{\xi}} d\nu + 2e^{-\frac{1}{\xi}}$$

$$\overset{(10.12)}{=} \sqrt{\pi\xi}\left(1 - \operatorname{erf}\left(\frac{1}{\sqrt{\xi}}\right)\right) + 1 + 2e^{-\frac{1}{\xi}}. \tag{10.22}$$

Let $x, y \in \mathbb{R} : |x - y| \leq \delta$, $\delta > 0$.

Then

$$(W_\xi f)(x) - (W_\xi f)(y)$$

$$\overset{(10.5)}{=} \frac{\sum_{\nu=-\infty}^{\infty} (f(x+\nu) - f(y+\nu)) e^{-\frac{\nu^2}{\xi}}}{\sqrt{\pi\xi}\left(1 - \operatorname{erf}\left(\frac{1}{\sqrt{\xi}}\right)\right) + 1}. \tag{10.23}$$

Hence

$$|(W_\xi f)(x) - (W_\xi f)(y)| \tag{10.24}$$

$$\leq \frac{\sum_{\nu=-\infty}^{\infty} |f(x+\nu) - f(y+\nu)| e^{-\frac{\nu^2}{\xi}}}{\sqrt{\pi\xi}\left(1 - \operatorname{erf}\left(\frac{1}{\sqrt{\xi}}\right)\right) + 1}$$

$$\leq \frac{\left(\sum_{\nu=-\infty}^{\infty} e^{-\frac{\nu^2}{\xi}}\right) \omega_1(f, |x-y|)}{\left(\sqrt{\pi\xi}\left(1 - \operatorname{erf}\left(\frac{1}{\sqrt{\xi}}\right)\right) + 1\right)} \tag{10.25}$$

$$\overset{((10.21), ((10.22))}{\leq} \frac{\left(\sqrt{\pi\xi}\left(1 - \operatorname{erf}\left(\frac{1}{\sqrt{\xi}}\right)\right) + 1 + 2e^{-\frac{1}{\xi}}\right) \omega_1(f, \delta)}{\sqrt{\pi\xi}\left(1 - \operatorname{erf}\left(\frac{1}{\sqrt{\xi}}\right)\right) + 1}$$

$$= \left[1 + \frac{2e^{-\frac{1}{\xi}}}{\sqrt{\pi\xi}\left(1 - \operatorname{erf}\left(\frac{1}{\sqrt{\xi}}\right)\right) + 1}\right] \omega_1(f, \delta), \tag{10.26}$$

proving the claim. ☐

We continue with the study of approximation properties of discrete Picard operators P_ξ.

Theorem 10.3. *Let $f \in C_U(\mathbb{R})$ or $f \in C_b(\mathbb{R})$, $x \in \mathbb{R}$, $0 < \xi \leq 1$. Then*

i)

$$|(P_\xi f)(x) - f(x)| \leq \frac{1}{\left(1 + 2\xi e^{-\frac{1}{\xi}}\right)}$$

$$\cdot \left[2 \left|f(x)\right| e^{-\frac{1}{\xi}} + \left(1 + 2e^{-\frac{1}{\xi}} \left(2\xi + 2 + \frac{1}{\xi}\right)\right) \omega_1 (f, \xi)\right], \tag{10.27}$$

and for $f \in C_U(\mathbb{R})$, as $\xi \to 0+$, we obtain $(P_\xi f)(x) \to f(x)$, pointwise,

ii) for $f \in C_b(\mathbb{R})$ we derive

$$\|P_\xi f - f\|_\infty \leq \frac{1}{\left(1 + 2\xi e^{-\frac{1}{\xi}}\right)}$$

$$\cdot \left[2 \|f\|_\infty e^{-\frac{1}{\xi}} + \left(1 + 2e^{-\frac{1}{\xi}} \left(2\xi + 2 + \frac{1}{\xi}\right)\right) \omega_1 (f, \xi)\right], \tag{10.28}$$

and for $f \in C_U(\mathbb{R}) \cap C_b(\mathbb{R})$, as $\xi \to 0+$, we get $P_\xi f \to f$ uniformly, i.e. $P_\xi \xrightarrow{u} I$.

Proof. We have that

$$\sum_{\nu=1}^{\infty} e^{-\frac{\nu}{\xi}} - e^{-\frac{1}{\xi}} \leq \int_1^\infty e^{-\frac{\nu}{\xi}} d\nu \leq \sum_{\nu=1}^{\infty} e^{-\frac{\nu}{\xi}}, \tag{10.29}$$

that is

$$0 \leq \sum_{\nu=1}^{\infty} e^{-\frac{\nu}{\xi}} - \int_1^\infty e^{-\frac{\nu}{\xi}} d\nu \leq e^{-\frac{1}{\xi}} \to 0, \text{ as } \xi \to 0+. \tag{10.30}$$

Notice that $\int_1^\infty e^{-\frac{\nu}{\xi}} d\nu = \xi e^{-\frac{1}{\xi}}$, consequently it holds

$$\sum_{\nu=-\infty}^{\infty} e^{-\frac{|\nu|}{\xi}} = 1 + 2\sum_{\nu=1}^{\infty} e^{-\frac{\nu}{\xi}} \leq 1 + 2e^{-\frac{1}{\xi}} (\xi + 1). \tag{10.31}$$

Also we have that

$$2\int_1^\infty e^{-\frac{\nu}{\xi}} d\nu + 1 = 2\xi e^{-\frac{1}{\xi}} + 1 \to 1, \text{ as } \xi \to 0+. \tag{10.32}$$

Similarly we find

$$\sum_{\nu=-\infty}^{\infty} \left(1 + \frac{|\nu|}{\sqrt{\xi}}\right) e^{-\frac{|\nu|}{\xi}} = 1 + 2\sum_{\nu=1}^{\infty} \left(1 + \frac{\nu}{\xi}\right) e^{-\frac{\nu}{\xi}}$$

$$\leq 1 + 2\left[\int_1^\infty \left(1 + \frac{\nu}{\xi}\right) e^{-\frac{\nu}{\xi}} d\nu + \left(1 + \frac{1}{\xi}\right) e^{-\frac{1}{\xi}}\right]$$

$$= 1 + 2e^{-\frac{1}{\xi}} \left(2\xi + 2 + \frac{1}{\xi}\right).$$

That is

$$\sum_{\nu=-\infty}^{\infty} \left(1 + \frac{|\nu|}{\sqrt{\xi}}\right) e^{-\frac{|\nu|}{\xi}} \leq 1 + 2e^{-\frac{1}{\xi}} \left(2\xi + 2 + \frac{1}{\xi}\right) \to 1, \text{ as } \xi \to 0+. \tag{10.33}$$

Also we have that

$$(P_\xi f)(x) \overset{(10.32)}{=} \frac{\sum_{\nu=-\infty}^{\infty} f(x+\nu) e^{-\frac{|\nu|}{\xi}}}{1+2\int_1^\infty e^{-\frac{\nu}{\xi}} d\nu}, \tag{10.34}$$

and it holds

$$(P_\xi f)(x) - f(x) = \tag{10.35}$$

$$\frac{2f(x)\left[\sum_{\nu=1}^{\infty} e^{-\frac{\nu}{\xi}} - \int_1^\infty e^{-\frac{\nu}{\xi}} d\nu\right] + \sum_{\nu=-\infty}^{\infty} (f(x+\nu) - f(x)) e^{-\frac{|\nu|}{\xi}}}{1+2\int_1^\infty e^{-\frac{\nu}{\xi}} d\nu}.$$

Therefore we obtain

$$|(P_\xi f)(x) - f(x)|$$

$$\overset{((10.30),\,(10.32),\,(10.35))}{\leq} \frac{2|f(x)| e^{-\frac{1}{\xi}} + \sum_{\nu=-\infty}^{\infty} |f(x+\nu) - f(x)| e^{-\frac{|\nu|}{\xi}}}{2\xi e^{-\frac{1}{\xi}} + 1} \tag{10.36}$$

$$\leq \frac{2|f(x)| e^{-\frac{1}{\xi}} + \sum_{\nu=-\infty}^{\infty} \omega_1(f, |\nu|) e^{-\frac{|\nu|}{\xi}}}{2\xi e^{-\frac{1}{\xi}} + 1}$$

$$\overset{(10.4)}{\leq} \frac{2|f(x)| e^{-\frac{1}{\xi}} + \omega_1(f,\xi) \sum_{\nu=-\infty}^{\infty} \left(1+\frac{|\nu|}{\xi}\right) e^{-\frac{|\nu|}{\xi}}}{2\xi e^{-\frac{1}{\xi}} + 1} \tag{10.37}$$

$$\overset{(10.33)}{\leq} \frac{2|f(x)| e^{-\frac{1}{\xi}} + \omega_1(f,\xi)\left(1+2e^{-\frac{1}{\xi}}\left(2\xi+2+\frac{1}{\xi}\right)\right)}{2\xi e^{-\frac{1}{\xi}} + 1}.$$

The proof of the theorem is now complete. □

Finally we show the global smoothness preservation property of operators P_ξ. So that the approximation of $P_\xi f$ to f is nice, smooth and fit.

Theorem 10.4. *Let $f \in C(\mathbb{R})$ with $\omega_1(f,\delta) < \infty$, $\forall\, \delta > 0$. Then*

$$\omega_1(P_\xi f, \delta) \leq \left(1 + \frac{2}{e^{\frac{1}{\xi}}\left(1+2\xi e^{-\frac{1}{\xi}}\right)}\right) \omega_1(f,\delta), \tag{10.38}$$

$\forall\, \delta > 0$, any $0 < \xi \leq 1$.

Proof. Let $x, y \in \mathbb{R} : |x-y| \leq \delta$, $\delta > 0$.

Then it holds

$$(P_\xi f)(x) - (P_\xi f)(y)$$

$$= \frac{\sum_{\nu=-\infty}^{\infty} (f(x+\nu) - f(y+\nu)) e^{-\frac{|\nu|}{\xi}}}{1+2\xi e^{-\frac{1}{\xi}}}. \tag{10.39}$$

Hence we have

$$|(P_{\xi}f)(x) - (P_{\xi}f)(y)| \tag{10.40}$$

$$\leq \frac{\sum_{\nu=-\infty}^{\infty} |f(x+\nu) - f(y+\nu)| \, e^{-\frac{|\nu|}{\xi}}}{1 + 2\xi e^{-\frac{1}{\xi}}}$$

$$\leq \frac{\left(\sum_{\nu=-\infty}^{\infty} e^{-\frac{|\nu|}{\xi}}\right) \omega_1(f, |x-y|)}{1 + 2\xi e^{-\frac{1}{\xi}}}$$

$$\overset{(10.31)}{\leq} \frac{\left(1 + 2e^{-\frac{1}{\xi}}(\xi+1)\right) \omega_1(f, \delta)}{1 + 2\xi e^{-\frac{1}{\xi}}}, \tag{10.41}$$

proving the claim. □

Bibliography

1. U. Abel, *Asymptotic expansions for the Favard operators and their left quasi-interpolants*, Stud. Univ. Babes-Bolyai Math. 56 (2) (2011), 199-206.
2. U. Abel, P. L. Butzer, *Complete asymptotic expansion for generalized Favard operators*, Constructive Approx. 35 (2012), 73-88.
3. O. Agratini, I.A. Rus, *Iterates of some bivariate approximation process via weakly Picard operators*, Nonlinear Anal. Forum 8 (2) (2003), 159-168.
4. G.A. Anastassiou, *On discrete Gauss–Weierstrass and Picard type operators*, Panamerican Math. J. 23 (2) (2013), 79-86.
5. G. Anastassiou, R. Mezei, *Approximation by singular integrals*, Cambridge Scientific Publishers, Cambridge, U.K., 2013.
6. C. Belingeri, P.E. Ricci, O. Villo, *A discretization of the Laplace and Gauss-Weierstrass operators*, Computers Math. Applic. 35 (8) (1998), 127-140.
7. R.A. DeVore, G.G. Lorentz, *Constructive Approximation*, Springer-Verlag, N.Y., Heidelberg, 1993.
8. I. Favard, *Sur les multiplicateurs d'interpolation*, J. Math. Pures Appl., IX, 23 (1944), 219-247.

Chapter 11

Approximation Theory by Interpolating Neural Networks

Here we introduce some general interpolating neural network operators in the univariate and multivariate cases. Initially we establish the interpolation property of the operators on functions. Then we derive the approximation properties of these operators on functions. We prove first the ordinary real quantitative pointwise and uniform convergences of these operators to the unit. Smoothness of functions is taken into consideration and speed of convergence improves dramatically. As extensions we consider also the fractional, fuzzy, fuzzy-fractional, fuzzy-random, complex and iterated cases. Furthermore we give Voronovskaya type asymptotic-expansions at all studied settings for the errors of related approximations. It follows [24].

11.1 Introduction

This chapter is mainly inspired by the great article of D. Costarelli [28], where he establishes interpolation and approximation properties of very specific neural network operators.

We present here the general related theory of similar general neural network operators. We expand to all possible directions.

The featured interpolation and approximation properties of our approximations are something very rare.

We mention next in very brief the initial D. Costarelli ([28]) theory.

We consider $C([a,b])$ the space of all continuous functions $f:[a,b]\rightarrow\mathbb{R}$, $a,b\in\mathbb{R}$, $a<b$. Let now $\sigma_R:\mathbb{R}\rightarrow[0,1]$ the ramp function defined by

$$\sigma_R(x) := \begin{cases} 0, \ x\leq -\frac{1}{2}, \\ 1, \ x\geq \frac{1}{2}, \\ x+\frac{1}{2}, \ \ -\frac{1}{2}<x<\frac{1}{2}. \end{cases} \tag{11.1}$$

The ramp function is a sigmoidal function $\sigma:\mathbb{R}\rightarrow\mathbb{R}$ which is measurable with $\lim_{x\rightarrow-\infty}\sigma(x)=0$ and $\lim_{x\rightarrow+\infty}\sigma(x)=1$. The last features arise in the theory of neural networks, where sigmoidal functions play the role of activation functions in the networks, see [39].

In [28], the author introduces

$$\Phi_R(x) := \sigma_R\left(x+\frac{1}{2}\right) - \sigma_R\left(x-\frac{1}{2}\right), \quad x \in \mathbb{R}. \tag{11.2}$$

The function $\Phi_R(x)$ has the properties: it is even, non-decreasing for $x < 0$ and non-increasing for $x \geq 0$, $\operatorname{supp}(\Phi_R) \subseteq [-1,1]$. Notice that $\Phi_R(\pm 1) = 0$.

Thus for $f : [a,b] \to \mathbb{R}$ a bounded and measurable function D. Costarelli [28], defines the neural network interpolation operator

$$F_n(f,x) := \frac{\sum_{k=0}^{n} f(x_k)\,\Phi_R\left(\frac{n(x-x_k)}{b-a}\right)}{\sum_{k=0}^{n} \Phi_R\left(\frac{n(x-x_k)}{b-a}\right)}, \quad x \in [a,b], \tag{11.3}$$

where the x_k's are the uniform spaced nodes defined by $x_k := a + kh$, $k = 0,1,...,n$, with $h := \frac{b-a}{n}$.

For a bounded measurable function f he proves

$$\|F_n(f)\|_\infty \leq \|f\|_\infty < +\infty, \tag{11.4}$$

where $\|f\|_\infty := \sup_{x\in[a,b]} |f(x)|$.

He also proves

Theorem 11.1. *([28]) Let $f : [a,b] \to \mathbb{R}$ a bounded measurable function and $n \in \mathbb{N}$. Then*

$$F_n(f,x_i) = f(x_i), \quad i = 0,1,...,n. \tag{11.5}$$

Theorem 11.2. *([28]) Let $f \in C([a,b])$. Then*

$$\|F_n(f) - f\|_\infty \leq 4\omega_1\left(f, \frac{b-a}{n}\right), \quad \forall\, n \in \mathbb{N}. \tag{11.6}$$

Above he uses

$$\omega_1(f,\delta) := \sup_{\substack{x,y:\\ |x-y|\leq\delta}} |f(x) - f(y)|, \quad 0 < \delta \leq b-a, \tag{11.7}$$

and if $\delta > b-a$, $\omega_1(f,\delta) := \omega_1(f,b-a)$, the first modulus of continuity.

D. Costarelli ([28]) gives also another specific example of interpolation neural network operators with the same properties as the F_n operators.

Denote by

$$M_s(x) := \frac{1}{(s-1)!}\sum_{i=0}^{s}(-1)^i\binom{s}{i}\left(\frac{s}{2}+x-i\right)_+^{s-1}, \quad x \in \mathbb{R}, \tag{11.8}$$

the B-spline of order $s \in \mathbb{N}$ ([26]), where $(x)_+ = \max\{x,0\}$, and $\operatorname{supp}(M_S) \subseteq \left[-\frac{s}{2}, \frac{s}{2}\right]$.

He defines ([28]) the sigmoidal functions

$$\sigma_{M_s}(x) := \int_{-\infty}^{x} M_s(t)\,dt, \quad x \in \mathbb{R}, \tag{11.9}$$

and the non-negative density functions:

$$\Phi_s(x) := \sigma_{M_s}\left(x + \frac{1}{2}\right) - \sigma_{M_s}\left(x - \frac{1}{2}\right), \quad x \in \mathbb{R},\ \forall\, s \in \mathbb{N}. \tag{11.10}$$

The functions Φ_s have the properties: even, non-decreasing for $x < 0$ and non-increasing for $x \geq 0$, $\operatorname{supp}(\Phi_s) \subseteq [-K_s, K_s] := \left[-\frac{(s+1)}{2}, \frac{(s+1)}{2}\right]$ and $\Phi_s\left(\frac{K_s}{2}\right) > 0$. Notice that $\Phi_s(\pm K_s) = 0$.

He ([28]) defines similarly the neural network operators

$$F_n^s(f, x) := \frac{\sum_{k=0}^{n} f(x_k)\, \Phi_s\left(K_s \frac{n(x - x_k)}{b-a}\right)}{\sum_{k=0}^{n} \Phi_s\left(K_s \frac{n(x - x_k)}{b-a}\right)}, \quad \forall\, x \in [a, b], \tag{11.11}$$

where $x_k := a + kh$, $k = 0, 1, ..., n$, and $h := \frac{b-a}{n}$.

Theorem 11.3. *([28]) Let $f : [a, b] \to \mathbb{R}$ a bounded and measurable function, $n \in \mathbb{N}$. Then*

$$F_n^s(f, x_k) = f(x_k), \quad k = 0, 1, ..., n, \quad s \in \mathbb{N}, \tag{11.12}$$

the interpolation property.

In addition, for $f \in C([a, b])$ we have

$$\|F_n^s(f) - f\|_\infty \leq \frac{2}{\Phi_s\left(\frac{K_s}{2}\right)} \omega_1\left(f, \frac{b-a}{n}\right), \quad \forall\, n, s \in \mathbb{N}. \tag{11.13}$$

Above the samples $f(x_k)$ can be viewed as the elements of the training set that can be used to train the normalized neural networks F_n, F_n^s. According to [28], the interpolation results show that the representation errors made by F_n, F_n^s on the elements of the training set are zero.

Furthermore the uniform approximation results, show the closeness property of neural network operators to well estimate elements outside the training set.

So our general theory presented in this chapter is the natural and complete outgrowth of [28] in very general diverse settings.

Other books and articles that inspired our work are: [12], [16], [17], [18], [19], [20], [21], [22], [23], [27], [37], [38].

The author was the first in 1997 to establish quantitative neural network approximations, see [1], [2], [3], [5], etc.

11.2 Main Results

11.2.1 *Neural Networks: Univariate Theory of Interpolation and Approximation*

We need

Definition 11.1. Let $B : \mathbb{R} \to \mathbb{R}_+$, be a bell-shaped function of compact support $[-T, T]$, $T > 0$. We assume it is even, non-decreasing for $x < 0$ and non-increasing for $x \geq 0$. Suppose also that $B(0) =: B^* > 0$ is the global maximum of B. The function B may have jump discontinuities and it is measurable. Assume further that $B(\pm T) = 0$.

Examples for B can be the hat function

$$\beta(x) := \begin{cases} 1 + x, & -1 \leq x \leq 0, \\ 1 - x, & 0 < x \leq 1, \\ 0, & \text{elsewhere}, \end{cases}$$

the function Φ_R, see (11.2), and the function Φ_s, see (11.10). Etc.

Definition 11.2. Let $f : [a, b] \to \mathbb{R}$, $a, b \in \mathbb{R}$, $a < b$, a bounded and measurable function, $n \in \mathbb{N}$, $h := \frac{b-a}{n}$, $x_k := a + kh$, $k = 0, 1, ..., n$, $x \in [a, b]$.

We define the interpolation neural network operator

$$H_n(f, x) := \frac{\sum_{k=0}^{n} f(x_k) B\left(\frac{Tn(x-x_k)}{b-a}\right)}{\sum_{k=0}^{n} B\left(\frac{Tn(x-x_k)}{b-a}\right)}. \tag{11.14}$$

We make

Remark 11.1. (on $H_n(f, x)$) We observe that

$$|H_n(f, x)| \leq \frac{\sum_{k=0}^{n} |f(x_k)| B\left(\frac{Tn(x-x_k)}{b-a}\right)}{\sum_{k=0}^{n} B\left(\frac{Tn(x-x_k)}{b-a}\right)} \leq \|f\|_\infty < +\infty. \tag{11.15}$$

That is

$$\|H_n(f)\|_\infty \leq \|f\|_\infty. \tag{11.16}$$

We make

Remark 11.2. Let $x \in [a, b]$, then $x_k \leq x \leq x_{k+1}$, for some $k \in \{0, 1, ..., n-1\}$, and $|x - x_k| \leq h$, $|x - x_{k+1}| \leq h$.

Notice that $B\left(\frac{Tn(x-x_k)}{b-a}\right) \neq 0$

$$\Leftrightarrow -T < \frac{Tn(x - x_k)}{b - a} < T$$

$$\Leftrightarrow -1 < \frac{n(x - x_k)}{b - a} < 1 \tag{11.17}$$
$$\Leftrightarrow -h < x - x_k < h$$
$$\Leftrightarrow |x - x_k| < h.$$

So when $x \in (x_k, x_{k+1})$, for some $k \in \{0, 1, ..., n-1\}$, we get both

$$B\left(\frac{Tn(x - x_k)}{b - a}\right), \ B\left(\frac{Tn(x - x_{k+1})}{b - a}\right) \neq 0.$$

When $x = x_k$, then

$$B\left(\frac{Tn(x_k - x_k)}{b - a}\right) = B(0) = B^* > 0,$$

and

$$B\left(\frac{Tn(x_k - x_{k+1})}{b - a}\right) = B(-T) = 0.$$

When $x = x_{k+1}$, then

$$B\left(\frac{Tn(x_{k+1} - x_k)}{b - a}\right) = B(T) = 0,$$

and

$$B\left(\frac{Tn(x_{k+1} - x_{k+1})}{b - a}\right) = B(0) = B^* > 0.$$

Clearly for any $x \in [x_k, x_{k+1}]$ we get that

$$B\left(\frac{Tn(x - x_i)}{b - a}\right) = 0, \ \text{ for all } i \neq k, k+1. \tag{11.18}$$

We make

Remark 11.3. For $x \in [a, b]$ we notice that

$$V(x) := \sum_{k=0}^{n} B\left(\frac{Tn(x - x_k)}{b - a}\right) = \sum_{k=0}^{n} B\left(\frac{Tn|x - x_k|}{b - a}\right) \geq B\left(\frac{Tn|x - x_i|}{b - a}\right), \tag{11.19}$$

where $i \in \{0, 1, ..., n\}$ is such that $|x - x_i| \leq \frac{h}{2}$. Thus

$$\frac{Tn|x - x_i|}{b - a} \leq \frac{Tnh}{2(b - a)} = \frac{T}{2}. \tag{11.20}$$

Therefore

$$B\left(\frac{Tn|x - x_i|}{b - a}\right) \geq B\left(\frac{T}{2}\right), \tag{11.21}$$

where $B\left(\frac{T}{2}\right) > 0$.

Thus $V(x) \geq B\left(\frac{T}{2}\right)$.

Consequently it holds

$$\frac{1}{V(x)} = \frac{1}{\sum_{k=0}^{n} B\left(\frac{Tn(x - x_k)}{b - a}\right)} \leq \frac{1}{B\left(\frac{T}{2}\right)}. \tag{11.22}$$

We state the interpolation result

Theorem 11.4. *Let $f : [a,b] \to \mathbb{R}$ be a bounded and measurable function. Then*

$$H_n(f, x_i) = f(x_i), \quad i = 0, 1, ..., n, \tag{11.23}$$

where $x_i := a + ih$, $h := \frac{b-a}{n}$, $n \in \mathbb{N}$.

Proof. Let $i \in \{0, 1, ..., n\}$ be fixed. When $k = i$, we have that

$$B\left(\frac{Tn(x_i - x_k)}{b-a}\right) = B(0) = B^* > 0. \tag{11.24}$$

But when $k \neq i$ we have

$$\frac{Tn|x_i - x_k|}{b-a} \geq \frac{Tnh}{b-a} = T, \tag{11.25}$$

hence

$$0 \leq B\left(\frac{Tn(x_i - x_k)}{b-a}\right) = B\left(\frac{Tn|x_i - x_k|}{b-a}\right) \leq B(T) = 0. \tag{11.26}$$

So we conclude that

$$B\left(\frac{Tn(x_i - x_k)}{b-a}\right) = \begin{Bmatrix} B^*, & i = k, \\ 0, & i \neq k \end{Bmatrix}, \tag{11.27}$$

for any $i, k = 0, 1, ..., n$.

By (11.27) we derive that

$$H_n(f, x_i) = \frac{f(x_i) B\left(\frac{Tn(x_i - x_i)}{b-a}\right)}{B\left(\frac{Tn(x_i - x_i)}{b-a}\right)} = \frac{f(x_i) B^*}{B^*} = f(x_i), \quad i = 0, 1, ..., n, \tag{11.28}$$

proving the claim. □

We state our first approximation result at Jackson speed of convergence $\frac{1}{n}$.

Theorem 11.5. *Let $f \in C([a,b])$. Then*

$$\|H_n(f) - f\|_\infty \leq \frac{2B^*}{B\left(\frac{T}{2}\right)} \omega_1\left(f, \frac{b-a}{n}\right), \quad \forall\, n \in \mathbb{N}. \tag{11.29}$$

Proof. Let $x \in [a,b]$, we can write

$$H_n(f,x) - f(x) = \frac{\sum_{k=0}^{n} f(x_k) B\left(\frac{Tn(x-x_k)}{b-a}\right)}{\sum_{k=0}^{n} B\left(\frac{Tn(x-x_k)}{b-a}\right)} - f(x)$$

$$= \frac{\sum_{k=0}^{n} f(x_k) B\left(\frac{Tn(x-x_k)}{b-a}\right) - f(x)\left(\sum_{k=0}^{n} B\left(\frac{Tn(x-x_k)}{b-a}\right)\right)}{V(x)}$$

$$= \frac{\sum_{k=0}^{n} \left(f\left(x_k\right) - f\left(x\right)\right) B\left(\frac{Tn(x-x_k)}{b-a}\right)}{V\left(x\right)}. \tag{11.30}$$

Therefore it holds

$$\left|H_n\left(f,x\right) - f\left(x\right)\right| \leq \frac{\sum_{k=0}^{n} \left|f\left(x_k\right) - f\left(x\right)\right| B\left(\frac{Tn(x-x_k)}{b-a}\right)}{V\left(x\right)}$$

$$\overset{(11.22)}{\leq} \frac{1}{B\left(\frac{T}{2}\right)} \left\{ \sum_{k=0}^{n} \left|f\left(x_k\right) - f\left(x\right)\right| B\left(\frac{Tn\left(x-x_k\right)}{b-a}\right) \right\} =: (*). \tag{11.31}$$

Let now $i \in \{0, 1, ..., n-1\}$ such that $x_i \leq x \leq x_{i+1}$. Hence

$$(*) = \frac{1}{B\left(\frac{T}{2}\right)} \left\{ \sum_{\substack{k=0 \\ k \neq i, i+1}}^{n} \left|f\left(x_k\right) - f\left(x\right)\right| B\left(\frac{Tn\left(x-x_k\right)}{b-a}\right) \right.$$

$$\left. + \left|f\left(x_i\right) - f\left(x\right)\right| B\left(\frac{Tn\left(x-x_i\right)}{b-a}\right) + \left|f\left(x_{i+1}\right) - f\left(x\right)\right| B\left(\frac{Tn\left(x-x_{i+1}\right)}{b-a}\right) \right\}$$

$$\leq \frac{1}{B\left(\frac{T}{2}\right)} \left\{0 + \omega_1\left(f,h\right) B^* + \omega_1\left(f,h\right) B^*\right\} = \frac{2B^*}{B\left(\frac{T}{2}\right)} \omega_1\left(f,h\right). \tag{11.32}$$

We derive for $f \in C\left([a,b]\right)$ that it holds

$$\left|H_n\left(f,x\right) - f\left(x\right)\right| \leq \frac{2B^*}{B\left(\frac{T}{2}\right)} \omega_1\left(f, \frac{b-a}{n}\right), \quad \forall\, x \in [a,b]. \tag{11.33}$$

The theorem now is proved. □

Taking into account the smoothness of f, we present the following high order approximation result.

Theorem 11.6. *Let $f \in C^N\left([a,b]\right)$, $N \in \mathbb{N}$, $x \in [a,b]$. Then*

i)

$$\left|H_n\left(f,x\right) - f\left(x\right)\right|$$

$$\leq \frac{2B^*}{B\left(\frac{T}{2}\right)} \left[\sum_{j=1}^{N} \frac{\left|f^{(j)}\left(x\right)\right|}{j!} \frac{(b-a)^j}{n^j} + \omega_1\left(f^{(N)}, \frac{b-a}{n}\right) \frac{(b-a)^N}{n^N N!} \right], \tag{11.34}$$

ii)

$$\left\|H_n\left(f\right) - f\right\|_\infty$$

$$\leq \frac{2B^*}{B\left(\frac{T}{2}\right)} \left[\sum_{j=1}^{N} \frac{\left\|f^{(j)}\right\|_\infty}{j!} \frac{(b-a)^j}{n^j} + \omega_1\left(f^{(N)}, \frac{b-a}{n}\right) \frac{(b-a)^N}{n^N N!} \right], \tag{11.35}$$

iii) assume more that $f^{(j)}(x) = 0$, $j = 1, ..., N$, where $x \in [a, b]$ is fixed, we get

$$|H_n(f, x) - f(x)| \leq \frac{2B^*}{B\left(\frac{T}{2}\right)} \omega_1\left(f^{(N)}, \frac{b-a}{n}\right) \frac{(b-a)^N}{n^N N!}, \tag{11.36}$$

a high speed $\frac{1}{n^{N+1}}$ pointwise convergence, and
iv)

$$\left| H_n(f, x) - f(x) - \sum_{j=1}^{N} \frac{f^{(j)}(x)}{j!} H_n\left((\cdot - x)^j, x\right) \right|$$
$$\leq \frac{2B^*}{B\left(\frac{T}{2}\right)} \omega_1\left(f^{(N)}, \frac{b-a}{n}\right) \frac{(b-a)^N}{n^N N!}. \tag{11.37}$$

Proof. Let $f \in C^N([a, b])$, $N \in \mathbb{N}$. Then

$$f(x_k) = \sum_{j=0}^{N} \frac{f^{(j)}(x)}{j!} (x_k - x)^j + \int_x^{x_k} \left(f^{(N)}(t) - f^{(N)}(x)\right) \frac{(x_k - t)^{N-1}}{(N-1)!} dt. \tag{11.38}$$

Hence it holds

$$\frac{f(x_k) B\left(\frac{Tn(x-x_k)}{b-a}\right)}{V(x)} = \sum_{j=0}^{N} \frac{f^{(j)}(x)}{j!} (x_k - x)^j \frac{B\left(\frac{Tn(x-x_k)}{b-a}\right)}{V(x)}$$
$$+ \frac{B\left(\frac{Tn(x-x_k)}{b-a}\right)}{V(x)} \int_x^{x_k} \left(f^{(N)}(t) - f^{(N)}(x)\right) \frac{(x_k - t)^{N-1}}{(N-1)!} dt. \tag{11.39}$$

Thus we can write

$$H_n(f, x) - f(x) = \frac{\sum_{k=0}^{n} f(x_k) B\left(\frac{Tn(x-x_k)}{b-a}\right)}{V(x)} - f(x)$$
$$= \sum_{j=1}^{N} \frac{f^{(j)}(x)}{j!} \frac{\sum_{k=0}^{n} (x_k - x)^j B\left(\frac{Tn(x-x_k)}{b-a}\right)}{V(x)}$$
$$+ \frac{\sum_{k=0}^{n} B\left(\frac{Tn(x-x_k)}{b-a}\right)}{V(x)} \int_x^{x_k} \left(f^{(N)}(t) - f^{(N)}(x)\right) \frac{(x_k - t)^{N-1}}{(N-1)!} dt. \tag{11.40}$$

Call

$$R_n(x) := \frac{\sum_{k=0}^{n} B\left(\frac{Tn(x-x_k)}{b-a}\right)}{V(x)} \int_x^{x_k} \left(f^{(N)}(t) - f^{(N)}(x)\right) \frac{(x_k - t)^{N-1}}{(N-1)!} dt. \tag{11.41}$$

Also call

$$\gamma\left(x, x_k\right) := \left| \int_x^{x_k} \left(f^{(N)}\left(t\right) - f^{(N)}\left(x\right) \right) \frac{\left(x_k - t\right)^{N-1}}{(N-1)!} dt \right| . \tag{11.42}$$

We distinguish the cases:

(i) Let $x \leq x_k$, then

$$\gamma\left(x, x_k\right) \leq \int_x^{x_k} \left| f^{(N)}\left(t\right) - f^{(N)}\left(x\right) \right| \frac{\left(x_k - t\right)^{N-1}}{(N-1)!} dt$$

$$\leq \omega_1\left(f^{(N)}, x_k - x \right) \frac{\left(x_k - x\right)^N}{N!}. \tag{11.43}$$

(ii) Let $x \geq x_k$, then

$$\gamma\left(x, x_k\right) = \left| \int_{x_k}^{x} \left(f^{(N)}\left(t\right) - f^{(N)}\left(x\right) \right) \frac{\left(t - x_k\right)^{N-1}}{(N-1)!} dt \right|$$

$$\leq \int_{x_k}^{x} \left| f^{(N)}\left(t\right) - f^{(N)}\left(x\right) \right| \frac{\left(t - x_k\right)^{N-1}}{(N-1)!} dt \leq \omega_1\left(f^{(N)}, x - x_k \right) \frac{\left(x - x_k\right)^N}{N!}. \tag{11.44}$$

We have found that

$$\gamma\left(x, x_k\right) \leq \omega_1\left(f^{(N)}, \left|x - x_k\right| \right) \frac{\left|x - x_k\right|^N}{N!}. \tag{11.45}$$

Therefore it holds

$$\left|R_n\left(x\right)\right| \leq \frac{\sum_{k=0}^{n} B\left(\frac{Tn(x-x_k)}{b-a}\right)}{V\left(x\right)} \omega_1\left(f^{(N)}, \left|x - x_k\right| \right) \frac{\left|x - x_k\right|^N}{N!} =: (*) . \tag{11.46}$$

Given that $x_k \leq x \leq x_{k+1}$, for some $k \in \{0, 1, ..., n-1\}$, we get

$$(*) = \frac{B\left(\frac{Tn(x-x_k)}{b-a}\right) \omega_1\left(f^{(N)}, \left|x - x_k\right| \right) \frac{\left|x-x_k\right|^N}{N!}}{V\left(x\right)}$$

$$+ \frac{B\left(\frac{Tn(x-x_{k+1})}{b-a}\right) \omega_1\left(f^{(N)}, \left|x - x_{k+1}\right| \right) \frac{\left|x-x_{k+1}\right|^N}{N!}}{V\left(x\right)}$$

$$\leq \frac{2B^* \omega_1\left(f^{(N)}, \frac{b-a}{n} \right) \frac{(b-a)^N}{n^N N!}}{B\left(\frac{T}{2}\right)}. \tag{11.47}$$

We have proved that

$$\left|R_n\left(x\right)\right| \leq \frac{2B^*}{B\left(\frac{T}{2}\right)} \omega_1\left(f^{(N)}, \frac{b-a}{n} \right) \frac{\left(b-a\right)^N}{n^N N!}. \tag{11.48}$$

Next we observe

$$\frac{\left|\sum_{k=0}^{n}(x_k-x)^j B\left(\frac{Tn(x-x_k)}{b-a}\right)\right|}{V(x)} \leq \frac{\sum_{k=0}^{n}|x_k-x|^j B\left(\frac{Tn(x-x_k)}{b-a}\right)}{V(x)}$$

$$\leq \frac{1}{B\left(\frac{T}{2}\right)}\left\{|x_k-x|^j B\left(\frac{Tn(x-x_k)}{b-a}\right) + |x_{k+1}-x|^j B\left(\frac{Tn(x-x_{k+1})}{b-a}\right)\right\}$$

$$\leq \frac{2B^* \frac{(b-a)^j}{n^j}}{B\left(\frac{T}{2}\right)}. \tag{11.49}$$

Therefore we derive

$$\left|\sum_{j=1}^{N}\frac{f^{(j)}(x)}{j!}\frac{\sum_{k=0}^{n}(x_k-x)^j B\left(\frac{Tn(x-x_k)}{b-a}\right)}{V(x)}\right|$$

$$\leq \frac{2B^*}{B\left(\frac{T}{2}\right)}\left(\sum_{j=1}^{N}\frac{\left|f^{(j)}(x)\right|}{j!}\frac{(b-a)^j}{n^j}\right). \tag{11.50}$$

Using (11.48) and (11.50) we derive (11.34)-(11.36).

Noticing that

$$\frac{\sum_{k=0}^{n}(x_k-x)^j B\left(\frac{Tn(x-x_k)}{b-a}\right)}{V(x)} = H_n\left((\cdot-x)^j, x\right), \tag{11.51}$$

we derive (11.37).

The theorem is proved. □

We present a related Voronovskaya type asymptotic expansion for the error of approximation.

Theorem 11.7. *Let $f \in C^N([a,b])$, $N \in \mathbb{N}$. Then*

$$H_n(f,x) - f(x) - \sum_{j=1}^{N-1}\frac{f^{(j)}(x)}{j!}H_n\left((\cdot-x)^j, x\right) = o\left(\frac{1}{n^{N-\varepsilon}}\right), \tag{11.52}$$

where $0 < \varepsilon \leq N$, $n \in \mathbb{N}$.

If $N = 1$, the sum above disappears.

Asymptotic expansion (11.52) implies

$$n^{N-\varepsilon}\left[H_n(f,x) - f(x) - \sum_{j=1}^{N-1}\frac{f^{(j)}(x)}{j!}H_n\left((\cdot-x)^j, x\right)\right] \to 0, \text{ as } n \to \infty, \tag{11.53}$$

$0 < \varepsilon \leq N$.

When $N = 1$, or $f^{(j)}(x) = 0$, $j = 1, ..., N-1$, then

$$n^{N-\varepsilon}\left[H_n(f,x) - f(x)\right] \to 0, \text{ as } n \to \infty,\ 0 < \varepsilon \leq N. \tag{11.54}$$

Proof. Let $x \in [a,b]$, then

$$f(x_k) = \sum_{j=0}^{N-1} \frac{f^{(j)}(x)}{j!}(x_k - x)^j + \int_x^{x_k} f^{(N)}(t) \frac{(x_k - t)^{N-1}}{(N-1)!} dt. \tag{11.55}$$

Let here $i \in \{0, 1, ..., n-1\}$ such that $x_i \leq x \leq x_{i+1}$.

Hence we have

$$\frac{f(x_k) B\left(\frac{Tn(x-x_k)}{b-a}\right)}{V(x)} = \sum_{j=0}^{N-1} \frac{f^{(j)}(x)}{j!}(x_k - x)^j \frac{B\left(\frac{Tn(x-x_k)}{b-a}\right)}{V(x)} \tag{11.56}$$

$$+ \frac{B\left(\frac{Tn(x-x_k)}{b-a}\right)}{V(x)} \int_x^{x_k} f^{(N)}(t) \frac{(x_k - t)^{N-1}}{(N-1)!} dt.$$

Thus it holds

$$H_n(f,x) - f(x) = \frac{\sum_{k=0}^{n} f(x_k) B\left(\frac{Tn(x-x_k)}{b-a}\right)}{V(x)} - f(x)$$

$$= \sum_{j=1}^{N-1} \frac{f^{(j)}(x)}{j!} \frac{\sum_{k=0}^{n} (x_k - x)^j B\left(\frac{Tn(x-x_k)}{b-a}\right)}{V(x)} \tag{11.57}$$

$$+ \frac{\sum_{k=0}^{n} B\left(\frac{Tn(x-x_k)}{b-a}\right)}{V(x)} \int_x^{x_k} f^{(N)}(t) \frac{(x_k - t)^{N-1}}{(N-1)!} dt.$$

Call

$$R(x) := \frac{\sum_{k=0}^{n} B\left(\frac{Tn(x-x_k)}{b-a}\right)}{V(x)} \int_x^{x_k} f^{(N)}(t) \frac{(x_k - t)^{N-1}}{(N-1)!} dt. \tag{11.58}$$

So that

$$H_n(f,x) - f(x) - \sum_{j=1}^{N-1} \frac{f^{(j)}(x)}{j!} H_n\left((\cdot - x)^j, x\right) = R(x). \tag{11.59}$$

Hence it holds

$$|R(x)| \leq \frac{\sum_{k=0}^{n} B\left(\frac{Tn(x-x_k)}{b-a}\right)}{B\left(\frac{T}{2}\right)} \left| \int_x^{x_k} f^{(N)}(t) \frac{(x_k - t)^{N-1}}{(N-1)!} dt \right| \leq (*). \tag{11.60}$$

But we find:

i) Let $x_k \geq x$. Then

$$\left| \int_x^{x_k} f^{(N)}(t) \frac{(x_k - t)^{N-1}}{(N-1)!} dt \right|$$

$$\leq \int_x^{x_k} \left| f^{(N)}(t) \right| \frac{(x_k - t)^{N-1}}{(N-1)!} dt \leq \left\| f^{(N)} \right\|_\infty \frac{(x_k - x)^N}{N!}. \tag{11.61}$$

ii) Let $x_k \leq x$. Then

$$\left| \int_x^{x_k} f^{(N)}(t) \frac{(x_k - t)^{N-1}}{(N-1)!} dt \right| = \left| \int_{x_k}^{x} f^{(N)}(t) \frac{(t - x_k)^{N-1}}{(N-1)!} dt \right|$$

$$\leq \int_{x_k}^{x} \left| f^{(N)}(t) \right| \frac{(t - x_k)^{N-1}}{(N-1)!} dt \leq \left\| f^{(N)} \right\|_\infty \frac{(x - x_k)^N}{N!}. \tag{11.62}$$

So in either case we have proved

$$\left| \int_x^{x_k} f^{(N)}(t) \frac{(x_k - t)^{N-1}}{(N-1)!} dt \right| \leq \left\| f^{(N)} \right\|_\infty \frac{|x - x_k|^N}{N!}. \tag{11.63}$$

Therefore we find

$$(*) \leq \frac{\sum_{k=0}^{n} B\left(\frac{Tn(x-x_k)}{b-a}\right)}{B\left(\frac{T}{2}\right)} \left\| f^{(N)} \right\|_\infty \frac{|x - x_k|^N}{N!}$$

$$\leq \frac{2B^*}{B\left(\frac{T}{2}\right)} \left\| f^{(N)} \right\|_\infty \frac{(b-a)^N}{N! n^N}. \tag{11.64}$$

We have proved that

$$|R(x)| \leq \frac{\psi}{n^N}, \tag{11.65}$$

where

$$\psi := \frac{2B^*}{B\left(\frac{T}{2}\right)} \frac{\left\| f^{(N)} \right\|_\infty (b-a)^N}{N!}. \tag{11.66}$$

Hence we derive

$$|R(x)| = O\left(\frac{1}{n^N}\right), \tag{11.67}$$

and

$$|R(x)| = o(1). \tag{11.68}$$

Letting $0 < \varepsilon \leq N$, we obtain

$$\frac{|R(x)|}{\left(\frac{1}{n^{N-\varepsilon}}\right)} \leq \frac{\psi}{n^\varepsilon} \to 0, \tag{11.69}$$

as $n \to \infty$. So that

$$|R(x)| = o\left(\frac{1}{n^{N-\varepsilon}}\right), \quad n \in \mathbb{N}, \tag{11.70}$$

proving the claim. □

We need

Definition 11.3. Let $\nu > 0$, $m = \lceil \nu \rceil$ ($\lceil \cdot \rceil$ is the ceiling of the number), $f \in AC^m([a,b])$ (space of functions f with $f^{(m-1)} \in AC([a,b])$, absolutely continuous functions). We call left Caputo fractional derivative (see [29], pp. 49-52, [32], [40]) the function

$$D_{*a}^{\nu} f(x) := \frac{1}{\Gamma(m-\nu)} \int_a^x (x-t)^{m-\nu-1} f^{(m)}(t)\, dt, \tag{11.71}$$

$\forall\, x \in [a,b]$, where Γ is the gamma function $\Gamma(\nu) := \int_0^{\infty} e^{-t} t^{\nu-1} dt$, $\nu > 0$.

We set $D_{*a}^{0} f(x) = f(x)$, $\forall\, x \in [a,b]$.

Lemma 11.1. *([8]) Let $\nu > 0$, $\nu \notin \mathbb{N}$, $m = \lceil \nu \rceil$, $f \in C^{m-1}([a,b])$ and $f^{(m)} \in L_{\infty}([a,b])$. Then $D_{*a}^{\nu} f(a) = 0$.*

Definition 11.4. (see also [9], [31], [32]) Let $f \in AC^m([a,b])$, $m = \lceil \nu \rceil$, $\nu > 0$. The right Caputo fractional derivative of order $\nu > 0$ is given by

$$D_{b-}^{\nu} f(x) := \frac{(-1)^m}{\Gamma(m-\nu)} \int_x^b (z-x)^{m-\nu-1} f^{(m)}(z)\, dz, \tag{11.72}$$

$\forall\, x \in [a,b]$. We set $D_{b-}^{0} f(x) = f(x)$.

Lemma 11.2. *([8]) Let $f \in C^{m-1}([a,b])$, $f^{(m)} \in L_{\infty}([a,b])$, $m = \lceil \nu \rceil$, $\nu > 0$. Then $D_{b-}^{\nu} f(b) = 0$.*

Convention 11.1. ([8]) We assume that

$$\begin{aligned} &D_{*x_0}^{\nu} f(x) = 0, \ \text{ for } x < x_0, \\ &\text{and} \\ &D_{x_0-}^{\nu} f(x) = 0, \ \text{ for } x > x_0, \end{aligned} \tag{11.73}$$

for all $x, x_0 \in [a,b]$.

We present the related fractional approximation result

Theorem 11.8. *Let $\beta > 0$, $N = \lceil \beta \rceil$, $\beta \notin \mathbb{N}$, $f \in AC^N([a,b])$, $f^{(N)} \in L_{\infty}([a,b])$. Then*

i)

$$|H_n(f,x) - f(x)| \le \frac{B^*}{B\left(\frac{T}{2}\right)} \left[2 \sum_{j=1}^{N-1} \frac{\left| f^{(j)}(x) \right|}{j!} \frac{(b-a)^j}{n^j} \right. \tag{11.74}$$

$$\left. + \frac{(b-a)^{\beta}}{\Gamma(\beta+1)\, n^{\beta}} \left[\omega_1\left(D_{x-}^{\beta} f, \frac{b-a}{n} \right) + \omega_1\left(D_{*x}^{\beta} f, \frac{b-a}{n} \right) \right] \right],$$

and

ii)

$$\|H_n(f) - f\|_\infty \leq \frac{B^*}{B\left(\frac{T}{2}\right)} \left[2 \sum_{j=1}^{N-1} \frac{\|f^{(j)}\|_\infty}{j!} \frac{(b-a)^j}{n^j} \right. \tag{11.75}$$

$$+ \frac{(b-a)^\beta}{\Gamma(\beta+1) n^\beta} \left[\sup_{x\in[a,b]} \omega_1 \left(D^\beta_{x-} f, \frac{b-a}{n} \right) \right.$$

$$\left. \left. + \sup_{x\in[a,b]} \omega_1 \left(D^\beta_{*x} f, \frac{b-a}{n} \right) \right] \right] < \infty.$$

Proof. Let fixed $x \in [a,b]$ with $x_i \leq x \leq x_{i+1}$, for some $i \in \{0,1,...,n-1\}$.

We have that

$$D^\beta_{x-} f(x) = D^\beta_{*x} = 0. \tag{11.76}$$

By Convention 11.1, $D^\beta_{*x} f(z) = 0$, for $z < x$; $D^\beta_{x-} f(z) = 0$, for $z > x$, all $x, z \in [a,b]$.

From [29], p. 54, we get by the left Caputo fractional Taylor formula that

$$f(x_k) = \sum_{j=0}^{N-1} \frac{f^{(j)}(x)}{j!} (x_k - x)^j \tag{11.77}$$

$$+ \frac{1}{\Gamma(\beta)} \int_x^{x_k} (x_k - J)^{\beta-1} \left(D^\beta_{*x} f(J) - D^\beta_{*x} f(x) \right) dJ,$$

for all $x \leq x_k \leq b$.

Also from [9], using the right Caputo fractional Taylor formula we get

$$f(x_k) = \sum_{j=0}^{N-1} \frac{f^{(j)}(x)}{j!} (x_k - x)^j \tag{11.78}$$

$$+ \frac{1}{\Gamma(\beta)} \int_{x_k}^{x} (J - x_k)^{\beta-1} \left(D^\beta_{x-} f(J) - D^\beta_{x-} f(x) \right) dJ,$$

for all $a \leq x_k \leq x$.

Hence it holds

$$\frac{f(x_k) B\left(\frac{Tn(x-x_k)}{b-a}\right)}{V(x)} = \sum_{j=0}^{N-1} \frac{f^{(j)}(x)}{j!} (x_k - x)^j \frac{B\left(\frac{Tn(x-x_k)}{b-a}\right)}{V(x)} \tag{11.79}$$

$$+ \frac{B\left(\frac{Tn(x-x_k)}{b-a}\right)}{V(x)} \frac{1}{\Gamma(\beta)} \int_x^{x_k} (x_k - J)^{\beta-1} \left(D^\beta_{*x} f(J) \right.$$

$$\left. - D^\beta_{*x} f(x) \right) dJ,$$

all $x \leq x_k \leq b$.

Also we have

$$\frac{f(x_k) B\left(\frac{Tn(x-x_k)}{b-a}\right)}{V(x)} = \sum_{j=0}^{N-1} \frac{f^{(j)}(x)}{j!} (x_k - x)^j \frac{B\left(\frac{Tn(x-x_k)}{b-a}\right)}{V(x)}$$

$$+\frac{B\left(\frac{Tn(x-x_k)}{b-a}\right)}{V(x)}\frac{1}{\Gamma(\beta)}\int_{x_k}^{x}(J-x_k)^{\beta-1}\left(D_{x-}^{\beta}f(J)-D_{x-}^{\beta}f(x)\right)dJ, \quad (11.80)$$

all $a \leq x_k \leq x$.

Hence we derive

$$\frac{\sum_{k=i+1}^{n} f(x_k)\,B\left(\frac{Tn(x-x_k)}{b-a}\right)}{V(x)} = \sum_{j=0}^{N-1}\frac{f^{(j)}(x)}{j!}\frac{\sum_{k=i+1}^{n}(x_k-x)^j B\left(\frac{Tn(x-x_k)}{b-a}\right)}{V(x)} + R_1, \quad (11.81)$$

where

$$R_1 := \frac{\sum_{k=i+1}^{n} B\left(\frac{Tn(x-x_k)}{b-a}\right)}{V(x)}\frac{1}{\Gamma(\beta)}\int_{x}^{x_k}(x_k-J)^{\beta-1}\left(D_{*x}^{\beta}f(J)-D_{*x}^{\beta}f(x)\right)dJ, \quad (11.82)$$

all $x \leq x_k \leq b$.

Also it holds

$$\frac{\sum_{k=0}^{i} f(x_k)\,B\left(\frac{Tn(x-x_k)}{b-a}\right)}{V(x)} = \sum_{j=0}^{N-1}\frac{f^{(j)}(x)}{j!}\frac{\sum_{k=0}^{i}(x_k-x)^j B\left(\frac{Tn(x-x_k)}{b-a}\right)}{V(x)} + R_2, \quad (11.83)$$

where

$$R_2 := \frac{\sum_{k=0}^{i} B\left(\frac{Tn(x-x_k)}{b-a}\right)}{V(x)}\frac{1}{\Gamma(\beta)}\int_{x_k}^{x}(J-x_k)^{\beta-1}\left(D_{x-}^{\beta}f(J)-D_{x-}^{\beta}f(x)\right)dJ, \quad (11.84)$$

all $a \leq x_k \leq x$.

Consequently, by adding (11.81) and (11.83), we obtain

$$H_n(f,x)-f(x) = \sum_{j=1}^{N-1}\frac{f^{(j)}(x)}{j!}\left(\frac{\sum_{k=0}^{n}(x_k-x)^j B\left(\frac{Tn(x-x_k)}{b-a}\right)}{V(x)}\right)+R_1+R_2. \quad (11.85)$$

Hence we find

$$|H_n(f,x)-f(x)| \leq \sum_{j=1}^{N-1}\frac{\left|f^{(j)}(x)\right|}{j!}\left(\frac{\sum_{k=0}^{n}|x_k-x|^j B\left(\frac{Tn(x-x_k)}{b-a}\right)}{V(x)}\right)+|R_1|+|R_2|$$

$$\leq \sum_{j=1}^{N-1}\frac{\left|f^{(j)}(x)\right|}{j!}\frac{2(b-a)^j B^*}{n^j B\left(\frac{T}{2}\right)}+|R_1|+|R_2|. \quad (11.86)$$

Next we estimate $|R_1|, |R_2|$.

We have that

$$|R_1| \leq \frac{\sum_{k=i+1}^{n} B\left(\frac{Tn(x-x_k)}{b-a}\right)}{B\left(\frac{T}{2}\right)} \frac{1}{\Gamma(\beta)} \int_x^{x_k} (x_k - J)^{\beta-1} \left|D_{*x}^{\beta} f(J) - D_{*x}^{\beta} f(x)\right| dJ \tag{11.87}$$

$$\leq \frac{\sum_{k=i+1}^{n} B\left(\frac{Tn(x-x_k)}{b-a}\right)}{B\left(\frac{T}{2}\right)} \frac{1}{\Gamma(\beta)} \omega_1\left(D_{*x}^{\beta} f, (x_k - x)\right) \left(\int_x^{x_k} (x_k - J)^{\beta-1} dJ\right)$$

$$= \frac{\sum_{k=i+1}^{n} B\left(\frac{Tn(x-x_k)}{b-a}\right)}{B\left(\frac{T}{2}\right)} \frac{1}{\Gamma(\beta)} \omega_1\left(D_{*x}^{\beta} f, x_k - x\right) \frac{(x_k - x)^{\beta}}{\beta}$$

$$= \frac{\sum_{k=i+1}^{n} B\left(\frac{Tn(x-x_k)}{b-a}\right)}{B\left(\frac{T}{2}\right) \Gamma(\beta+1)} (x_k - x)^{\beta} \omega_1\left(D_{*x}^{\beta} f, x_k - x\right) \tag{11.88}$$

$$\leq \frac{B\left(\frac{Tn(x-x_{i+1})}{b-a}\right)}{B\left(\frac{T}{2}\right) \Gamma(\beta+1)} \frac{(b-a)^{\beta}}{n^{\beta}} \omega_1\left(D_{*x}^{\beta} f, \frac{b-a}{n}\right). \tag{11.89}$$

We have proved that

$$|R_1| \leq \frac{B^*}{B\left(\frac{T}{2}\right) \Gamma(\beta+1)} \frac{(b-a)^{\beta}}{n^{\beta}} \omega_1\left(D_{*x}^{\beta} f, \frac{b-a}{n}\right). \tag{11.90}$$

Furthermore we observe that

$$|R_2| \leq \frac{\sum_{k=0}^{i} B\left(\frac{Tn(x-x_k)}{b-a}\right)}{B\left(\frac{T}{2}\right)} \frac{1}{\Gamma(\beta)} \left(\int_{x_k}^{x} (J - x_k)^{\beta-1} \left|D_{x-}^{\beta} f(J) - D_{x-}^{\beta} f(x)\right| dJ\right) \tag{11.91}$$

$$\leq \frac{\sum_{k=0}^{i} B\left(\frac{Tn(x-x_k)}{b-a}\right)}{B\left(\frac{T}{2}\right)} \frac{1}{\Gamma(\beta)} \omega_1\left(D_{x-}^{\beta} f, x - x_k\right) \frac{(x - x_k)^{\beta}}{\beta}$$

$$= \frac{B\left(\frac{Tn(x-x_i)}{b-a}\right)}{B\left(\frac{T}{2}\right)} \frac{1}{\Gamma(\beta+1)} (x - x_i)^{\beta} \omega_1\left(D_{x-}^{\beta} f, x - x_i\right) \tag{11.92}$$

$$\leq \frac{B^*}{B\left(\frac{T}{2}\right)} \frac{1}{\Gamma(\beta+1)} \frac{(b-a)^{\beta}}{n^{\beta}} \omega_1\left(D_{x-}^{\beta} f, \frac{b-a}{n}\right).$$

That is we have proved

$$|R_2| \leq \frac{B^*}{B\left(\frac{T}{2}\right) \Gamma(\beta+1)} \frac{(b-a)^{\beta}}{n^{\beta}} \omega_1\left(D_{x-}^{\beta} f, \frac{b-a}{n}\right). \tag{11.93}$$

Thus

$$|R_1| + |R_2| \leq \frac{B^* (b-a)^{\beta}}{B\left(\frac{T}{2}\right) \Gamma(\beta+1) n^{\beta}} \left[\omega_1 \left(D_{x-}^{\beta} f, \frac{b-a}{n}\right) + \omega_1 \left(D_{*x}^{\beta} f, \frac{b-a}{n}\right)\right]. \tag{11.94}$$

So by using (11.86) and (11.94) we obtain (11.74), which implies (11.75).

Next we justify that the right hand side of (11.75) is finite.

We have

$$\left(D_{*x}^{\beta} f\right)(t) = \frac{1}{\Gamma(N-\beta)} \int_x^t (t-z)^{N-\beta-1} f^{(N)}(z)\, dz, \quad x \leq t \leq b. \tag{11.95}$$

Hence

$$\left|D_{*x}^{\beta} f(t)\right| \leq \frac{\left\|f^{(N)}\right\|_{\infty}}{\Gamma(N-\beta+1)} (b-a)^{N-\beta}, \quad x \leq t \leq b. \tag{11.96}$$

Thus

$$\left\|D_{*x}^{\beta} f\right\|_{\infty} \leq \frac{\left\|f^{(N)}\right\|_{\infty}}{\Gamma(N-\beta+1)} (b-a)^{N-\beta}. \tag{11.97}$$

Similarly

$$D_{x-}^{\beta} f(t) = \frac{(-1)^N}{\Gamma(N-\beta)} \int_t^x (z-t)^{N-\beta-1} f^{(N)}(z)\, dz, \quad \text{all } a \leq t \leq x. \tag{11.98}$$

Hence

$$\left|D_{x-}^{\beta} f(t)\right| \leq \frac{\left\|f^{(N)}\right\|_{\infty}}{\Gamma(N-\beta+1)} (b-a)^{N-\beta}, \quad a \leq t \leq x. \tag{11.99}$$

Thus

$$\left\|D_{x-}^{\beta} f\right\|_{\infty} \leq \frac{\left\|f^{(N)}\right\|_{\infty}}{\Gamma(N-\beta+1)} (b-a)^{N-\beta}. \tag{11.100}$$

Consequently (for $\delta > 0$)

$$\omega_1\left(D_{x-}^{\beta} f, \delta\right) = \sup_{\substack{z_1, z_2 \\ |z_1 - z_2| \leq \delta}} \left|D_{x-}^{\beta} f(z_1) - D_{x-}^{\beta} f(z_2)\right| \leq$$

$$\sup_{\substack{z_1, z_2 \\ |z_1 - z_2| \leq \delta}} \left\{\left|D_{x-}^{\beta} f(z_1)\right| + \left|D_{x-}^{\beta} f(z_2)\right|\right\} \leq 2 \left\|D_{x-}^{\beta} f\right\|_{\infty}$$

$$\leq \frac{2\left\|f^{(N)}\right\|_{\infty}}{\Gamma(N-\beta+1)} (b-a)^{N-\beta} < +\infty. \tag{11.101}$$

Hence it holds

$$\omega_1\left(D_{x-}^{\beta} f, \delta\right) \leq \frac{2\left\|f^{(N)}\right\|_{\infty}}{\Gamma(N-\beta+1)} (b-a)^{N-\beta} < +\infty. \tag{11.102}$$

Therefore

$$\sup_{x \in [a,b]} \omega_1\left(D_{x-}^{\beta} f, \delta\right) \leq \frac{2\left\|f^{(N)}\right\|_{\infty}}{\Gamma(N-\beta+1)} (b-a)^{N-\beta} < +\infty, \tag{11.103}$$

and, similarly, we get

$$\sup_{x \in [a,b]} \omega_1\left(D_{*x}^{\beta} f, \delta\right) \leq \frac{2\left\|f^{(N)}\right\|_{\infty}}{\Gamma(N-\beta+1)} (b-a)^{N-\beta} < +\infty. \tag{11.104}$$

The proof of the theorem now is complete. □

Corollary 11.1. *(to Theorem 11.8) All as in Theorem 11.8. Additionally assume that* $f^{(j)}(x) = 0$, $j = 1, ..., N-1$. *Then*

$$|H_n(f,x) - f(x)| \leq \frac{B^*}{B\left(\frac{T}{2}\right)} \frac{(b-a)^\beta}{\Gamma(\beta+1) n^\beta} \cdot \tag{11.105}$$

$$\left[\omega_1\left(D_{x-}^\beta f, \frac{b-a}{n}\right) + \omega_1\left(D_{*x}^\beta f, \frac{b-a}{n}\right)\right].$$

In the last we have the high speed of pointwise convergence at $\frac{1}{n^{\beta+1}}$.

A fractional Voronovskaya type asymptotic expansion follows.

Theorem 11.9. *Let* $\beta > 0$, $N = \lceil \beta \rceil$, $\beta \notin \mathbb{N}$, $f \in AC^N([a,b])$, $f^{(N)} \in L_\infty([a,b])$. *Then*

$$H_n(f,x) - f(x) - \sum_{j=1}^{N-1} \frac{f^{(j)}(x)}{j!} H_n\left((\cdot - x)^j, x\right) = o\left(\frac{1}{n^{\beta-\varepsilon}}\right), \tag{11.106}$$

where $0 < \varepsilon \leq \beta$, $n \in \mathbb{N}$.

If $N = 1$, *the sum above disappears.*

Asymptotic expansion (11.106) implies

$$n^{\beta-\varepsilon}\left[H_n(f,x) - f(x) - \sum_{j=1}^{N-1} \frac{f^{(j)}(x)}{j!} H_n\left((\cdot - x)^j, x\right)\right] \to 0, \tag{11.107}$$

as $n \to \infty$, $0 < \varepsilon \leq \beta$.

When $N = 1$, *or* $f^{(j)}(x) = 0$, $j = 1, ..., N-1$, *then*

$$n^{\beta-\varepsilon}\left[H_n(f,x) - f(x)\right] \to 0, \tag{11.108}$$

as $n \to \infty$, $0 < \varepsilon \leq \beta$.

Of great interest is the case $\beta = \frac{1}{2}$.

Proof. From [29], p. 54, we get by the left Caputo fractional Taylor formula that

$$f(x_k) = \sum_{j=0}^{N-1} \frac{f^{(j)}(x)}{j!}(x_k - x)^j + \frac{1}{\Gamma(\beta)} \int_x^{x_k} (x_k - J)^{\beta-1} D_{*x}^\beta f(J)\, dJ, \tag{11.109}$$

for all $x \leq x_k \leq b$.

Also from [9], using the right Caputo fractional Taylor formula we get

$$f(x_k) = \sum_{j=0}^{N-1} \frac{f^{(j)}(x)}{j!}(x_k - x)^j + \frac{1}{\Gamma(\beta)} \int_{x_k}^{x} (J - x_k)^{\beta-1} D_{x-}^\beta f(J)\, dJ, \tag{11.110}$$

for all $a \leq x_k \leq x$.

Hence

$$\frac{f(x_k) B\left(\frac{Tn(x-x_k)}{b-a}\right)}{V(x)} = \sum_{j=0}^{N-1} \frac{f^{(j)}(x)}{j!}(x_k - x)^j \frac{B\left(\frac{Tn(x-x_k)}{b-a}\right)}{V(x)} \tag{11.111}$$

$$+\frac{B\left(\frac{Tn(x-x_k)}{b-a}\right)}{V(x)}\frac{1}{\Gamma(\beta)}\int_x^{x_k}(x_k-J)^{\beta-1}D_{*x}^{\beta}f(J)\,dJ,$$

all $x \leq x_k \leq b$.

Also we have

$$\frac{f(x_k)B\left(\frac{Tn(x-x_k)}{b-a}\right)}{V(x)}=\sum_{j=0}^{N-1}\frac{f^{(j)}(x)}{j!}(x_k-x)^j\frac{B\left(\frac{Tn(x-x_k)}{b-a}\right)}{V(x)}$$
$$+\frac{B\left(\frac{Tn(x-x_k)}{b-a}\right)}{V(x)}\frac{1}{\Gamma(\beta)}\int_{x_k}^{x}(J-x_k)^{\beta-1}D_{x-}^{\beta}f(J)\,dJ, \tag{11.112}$$

all $a \leq x_k \leq x$.

Hence $x \in [a,b]$ is fixed such that $x_i \leq x \leq x_{i+1}$, for some $i \in \{0,1,...,n-1\}$.

Hence it holds

$$\frac{\sum_{k=i+1}^{n}f(x_k)B\left(\frac{Tn(x-x_k)}{b-a}\right)}{V(x)}=\sum_{j=0}^{N-1}\frac{f^{(j)}(x)}{j!}\frac{\sum_{k=i+1}^{n}(x_k-x)^jB\left(\frac{Tn(x-x_k)}{b-a}\right)}{V(x)}+R_1, \tag{11.113}$$

where

$$R_1:=\frac{\sum_{k=i+1}^{n}B\left(\frac{Tn(x-x_k)}{b-a}\right)}{V(x)}\frac{1}{\Gamma(\beta)}\int_x^{x_k}(x_k-J)^{\beta-1}D_{*x}^{\beta}f(J)\,dJ, \tag{11.114}$$

all $x \leq x_k \leq b$.

Also it holds

$$\frac{\sum_{k=0}^{i}f(x_k)B\left(\frac{Tn(x-x_k)}{b-a}\right)}{V(x)}=\sum_{j=0}^{N-1}\frac{f^{(j)}(x)}{j!}\frac{\sum_{k=0}^{i}(x_k-x)^jB\left(\frac{Tn(x-x_k)}{b-a}\right)}{V(x)}+R_2, \tag{11.115}$$

where

$$R_2:=\frac{\sum_{k=0}^{i}B\left(\frac{Tn(x-x_k)}{b-a}\right)}{V(x)}\frac{1}{\Gamma(\beta)}\int_{x_k}^{x}(J-x_k)^{\beta-1}D_{x-}^{\beta}f(J)\,dJ, \tag{11.116}$$

all $a \leq x_k \leq x$.

Hence we get

$$H_n(f,x)-f(x)-\sum_{j=1}^{N-1}\frac{f^{(j)}(x)}{j!}\left(\frac{\sum_{k=0}^{n}(x_k-x)^jB\left(\frac{Tn(x-x_k)}{b-a}\right)}{V(x)}\right)=R_1+R_2. \tag{11.117}$$

Notice also that for any $x \in [a,b]$, by (11.97) and (11.100), we have

$$\left\{ \left\| D_{*x}^{\beta} f \right\|_{\infty}, \left\| D_{x-}^{\beta} f \right\|_{\infty} \right\} \leq \frac{\left\| f^{(N)} \right\|_{\infty}}{\Gamma(N-\beta+1)} (b-a)^{N-\beta} =: M, \tag{11.118}$$

with $M > 0$.

That is we find

$$H_n(f,x) - f(x) - \sum_{j=1}^{N-1} \frac{f^{(j)}(x)}{j!} H_n\left((\cdot - x)^j, x \right) = R_1 + R_2. \tag{11.119}$$

Notice that

$$|R_1| \leq M \frac{\sum_{k=i+1}^{n} B\left(\frac{Tn(x-x_k)}{b-a} \right)}{V(x)} \frac{1}{\Gamma(\beta+1)} (x_k - x)^{\beta} \tag{11.120}$$

$$\leq \frac{MB^*}{B\left(\frac{T}{2}\right)\Gamma(\beta+1)} \frac{(b-a)^{\beta}}{n^{\beta}},$$

that is

$$|R_1| \leq \frac{MB^*(b-a)^{\beta}}{B\left(\frac{T}{2}\right)\Gamma(\beta+1)n^{\beta}}. \tag{11.121}$$

Similarly we have

$$|R_2| \leq M \frac{\sum_{k=0}^{i} B\left(\frac{Tn(x-x_k)}{b-a} \right)}{B\left(\frac{T}{2}\right)} \frac{1}{\Gamma(\beta+1)} (x - x_k)^{\beta}$$

$$\leq \frac{MB^*}{B\left(\frac{T}{2}\right)\Gamma(\beta+1)} \frac{(b-a)^{\beta}}{n^{\beta}}. \tag{11.122}$$

Hence

$$|R_2| \leq \frac{MB^*(b-a)^{\beta}}{B\left(\frac{T}{2}\right)\Gamma(\beta+1)n^{\beta}}. \tag{11.123}$$

Therefore it holds

$$|R_1 + R_2| \leq |R_1| + |R_2| \leq \frac{\Phi}{n^{\beta}}, \tag{11.124}$$

where

$$\Phi := \frac{2MB^*(b-a)^{\beta}}{B\left(\frac{T}{2}\right)\Gamma(\beta+1)}. \tag{11.125}$$

Thus

$$|R_1 + R_2| = O\left(\frac{1}{n^{\beta}} \right), \tag{11.126}$$

and

$$|R_1 + R_2| = o(1).$$

Letting $0 < \varepsilon \leq \beta$, we derive

$$\frac{|R_1 + R_2|}{\left(\frac{1}{n^{\beta-\varepsilon}} \right)} \leq \frac{\Phi}{n^{\varepsilon}} \to 0, \tag{11.127}$$

as $n \to \infty$. So that

$$|R_1 + R_2| = o\left(\frac{1}{n^{\beta-\varepsilon}} \right), \quad n \in \mathbb{N}, \tag{11.128}$$

proving the claim. ∎

11.2.2 *Neural Networks: Multivariate Theory of Interpolation and Approximation*

We need

Definition 11.5. Consider the d-dimensional bell-shaped function $E : \mathbb{R}^d \to \mathbb{R}_+$ $(d \in \mathbb{N})$ with the property for all $i = 1, ..., d$, $\mathbb{R} \ni t \longrightarrow E(x_1, ..., t, ..., x_d)$ is a bell-shaped function, as in Definition 11.1, where $x = (x_1, ..., x_d) \in \mathbb{R}^d$ is arbitrary.

More precisely here E is of compact support $K := \prod_{i=1}^{d} [-T_i, T_i]$, $T_i > 0$ and it may have jump discontinuities there, also it holds

$$E(x_1, ..., \pm T_i, ..., x_d) = 0, \tag{11.129}$$

for any $i = 1, ..., d$, all $(x_1, ..., x_d) \in \mathbb{R}^d$.

Furthermore assume that $E(0, ..., 0) =: E^* > 0$ is the global maximum of E, also E is assumed to be measurable. That is $E(x_1, ..., t, ..., x_d)$ in t is even, non-decreasing for $t < 0$ and non-increasing for $t \geq 0$.

Clearly it holds

$$E(\pm x_1, ..., \pm x_d) = E(|x_1|, ..., |x_d|). \tag{11.130}$$

Also it is $E(x_1, ..., 0, ..., x_d) =: E^*(x_1, ..., x_{i-1}, x_{i+1}, ..., x_d) > 0$, for all $i = 1, ..., d$, for any $(x_1, ..., x_d) \in \prod_{i=1}^{d} (-T_i, T_i)$.

Examples: $\prod_{i=1}^{d} \beta(x_i)$, $\prod_{i=1}^{d} \Phi_R(x_i)$, $\prod_{i=1}^{d} \Phi_s(x_i)$, etc.

Definition 11.6. Let $f : \prod_{i=1}^{d} [a_i, b_i] \to \mathbb{R}$ be a bounded and measurable function, $a_i < b_i$, $n \in \mathbb{N}$, $h_i := \frac{b_i - a_i}{n}$, $x_{k_i i} := a_i + k_i h_i$, $k_i = 0, 1, ..., n$, $i = 1, ..., d$, $x = (x_1, ..., x_d) \in \prod_{i=1}^{d} [a_i, b_i]$.

Next we define the multivariate interpolation neural network operator:

$$M_n(f, x) := M_n(f, x_1, ..., x_d)$$

$$:= \frac{\sum_{k_1=0}^{n} \cdots \sum_{k_d=0}^{n} f(x_{k_1 1}, ..., x_{k_d d}) E\left(\frac{T_1 n(x_1 - x_{k_1 1})}{b_1 - a_1}, ..., \frac{T_d n(x_d - x_{k_d d})}{b_d - a_d}\right)}{\sum_{k_1=0}^{n} \cdots \sum_{k_d=0}^{n} E\left(\frac{T_1 n(x_1 - x_{k_1 1})}{b_1 - a_1}, ..., \frac{T_d n(x_d - x_{k_d d})}{b_d - a_d}\right)}. \tag{11.131}$$

Remark 11.4. Trivially we get that

$$|M_n(f, x)| \leq \|f\|_\infty < +\infty, \tag{11.132}$$

and

$$\|M_n(f)\|_\infty \leq \|f\|_\infty < +\infty. \tag{11.133}$$

Remark 11.5. Let now $x_{k_i i} < x_i < x_{(k_i+1)i}$, for all $i = 1, ..., d$, for some $(k_1, ..., k_d) \in \{0, 1, ..., n-1\}^d$. Thus $|x_i - x_{k_i i}| < h_i$ and $\left|x_i - x_{(k_i+1)i}\right| < h_i$, for all $i = 1, ..., d$, for some $(k_1, ..., k_d) \in \{0, 1, ..., n-1\}^d$.

Remark 11.6. Notice next that be given $(x_1, ..., x_d) \in \mathbb{R}^d$ and

$$E\left(\frac{T_1 n (x_1 - x_{k_1 1})}{b_1 - a_1}, ..., \frac{T_d n (x_d - x_{k_d d})}{b_d - a_d}\right) > 0, \tag{11.134}$$

for some $(k_1, ..., k_d) \in \{0, 1, ..., n\}^d$, $\Leftrightarrow$ simultaneously it holds

$$-T_i < \frac{T_i n (x_i - x_{k_i i})}{b_i - a_i} < T_i,$$

for all $i = 1, ..., d$, for some $(k_1, ..., k_d) \in \{0, 1, ..., n\}^d$, $\Leftrightarrow$

$$-1 < \frac{n (x_i - x_{k_i i})}{b_i - a_i} < 1, \tag{11.135}$$

for all $i = 1, ..., d$, for some $(k_1, ..., k_d) \in \{0, 1, ..., n\}^d$, $\Leftrightarrow$

$$-h_i < x_i - x_{k_i i} < h_i,$$

for all $i = 1, ..., d$, for some $(k_1, ..., k_d) \in \{0, 1, ..., n\}^d$, $\Leftrightarrow$

$$|x_i - x_{k_i i}| < h_i, \tag{11.136}$$

for all $i = 1, ..., d$, for some $(k_1, ..., k_d) \in \{0, 1, ..., n\}^d$.

Thus, when $x \in \prod_{i=1}^{d} \left[x_{k_i i}, x_{(k_i+1)i}\right)$, for some $(k_1, ..., k_d) \in \{0, 1, ..., n-1\}^d$, we get that

$$E\left(\frac{T_1 n (x_1 - x_{k_1 1})}{b_1 - a_1}, ..., \frac{T_d n (x_d - x_{k_d d})}{b_d - a_d}\right) > 0. \tag{11.137}$$

Remark 11.7. Notice that $\left(x \in \prod_{i=1}^{d} [a_i, b_i]\right)$

$$W := \sum_{k_1=0}^{n} ... \sum_{k_d=0}^{n} E\left(\frac{T_1 n (x_1 - x_{k_1 1})}{b_1 - a_1}, ..., \frac{T_d n (x_d - x_{k_d d})}{b_d - a_d}\right)$$

$$= \sum_{k_1=0}^{n} ... \sum_{k_d=0}^{n} E\left(\frac{T_1 n |x_1 - x_{k_1 1}|}{b_1 - a_1}, ..., \frac{T_d n |x_d - x_{k_d d}|}{b_d - a_d}\right)$$

$$\geq E\left(\frac{T_1 n |x_1 - x_{k_1 1}|}{b_1 - a_1}, ..., \frac{T_d n |x_d - x_{k_d d}|}{b_d - a_d}\right), \tag{11.138}$$

the last inequality is chosen for suitable x_i and $x_{k_i i}$, for all $i = 1, ..., d$, and for some $(k_1, ..., k_d) \in \{0, 1, ..., n\}^d$, such that $|x_i - x_{k_i i}| \leq \frac{h_i}{2}$.

Thus

$$\frac{T_i n |x_i - x_{k_i i}|}{b_i - a_i} \leq \frac{T_i n h_i}{2 (b_i - a_i)} = \frac{T_i}{2}, \tag{11.139}$$

all $i = 1, ..., d$.

Therefore it holds

$$E\left(\frac{T_1 n\left|x_1 - x_{k_1 1}\right|}{b_1 - a_1}, \frac{T_2 n\left|x_2 - x_{k_2 2}\right|}{b_2 - a_2}, ..., \frac{T_d n\left|x_d - x_{k_d d}\right|}{b_d - a_d}\right)$$

$$\geq E\left(\frac{T_1}{2}, \frac{T_2 n\left|x_2 - x_{k_2 2}\right|}{b_2 - a_2}, ..., \frac{T_d n\left|x_d - x_{k_d d}\right|}{b_d - a_d}\right)$$

$$\geq E\left(\frac{T_1}{2}, \frac{T_2}{2}, ..., \frac{T_d n\left|x_d - x_{k_d d}\right|}{b_d - a_d}\right) \geq ... \geq E\left(\frac{T_1}{2}, \frac{T_2}{2}, ..., \frac{T_d}{2}\right) > 0. \quad (11.140)$$

Hence we have

$$\frac{1}{W} = \frac{1}{\sum_{k_1=0}^{n} ... \sum_{k_d=0}^{n} E\left(\frac{T_1 n\left(x_1 - x_{k_1 1}\right)}{b_1 - a_1}, ..., \frac{T_d n\left(x_d - x_{k_d d}\right)}{b_d - a_d}\right)} \leq \frac{1}{E\left(\frac{T_1}{2}, ..., \frac{T_d}{2}\right)}. \quad (11.141)$$

Remark 11.8. Let all $x_i = x_{k_i i}$, $i = 1, ..., d$, for some $(k_1, ..., k_d) \in \{0, 1, ..., n\}^d$. Then

$$E\left(\frac{T_1 n\left(x_1 - x_{k_1 1}\right)}{b_1 - a_1}, ..., \frac{T_d n\left(x_d - x_{k_d d}\right)}{b_d - a_d}\right) = E\left(0, ..., 0\right) = E^* > 0. \quad (11.142)$$

Let next $\left|x_{k_i i} - x_{(k_i + j_i) i}\right| \geq h_i$, for some $i \in \{1, ..., d\}$, where $j_i \geq 1$ integer, and $k_i, k_i + j_i \in \{0, 1, ..., n\}$.

Then

$$\frac{T_i n\left|x_{k_i i} - x_{(k_i + j_i) i}\right|}{b_i - a_i} \geq \frac{T_i n h_i}{b_i - a_i} = T_i, \quad \text{for some } i = 1, ..., d. \quad (11.143)$$

Hence

$$0 \leq E\left(\frac{T_1 n\left|x_1 - x_{k_1 1}\right|}{b_1 - a_1}, ..., \frac{T_i n\left|x_{(k_i + j_i) i} - x_{k_i i}\right|}{b_i - a_i}, ..., \frac{T_d n\left|x_d - x_{k_d d}\right|}{b_d - a_d}\right)$$

$$\leq E\left(\frac{T_1 n\left|x_1 - x_{k_1 1}\right|}{b_1 - a_1}, ..., T_i, ..., \frac{T_d n\left|x_d - x_{k_d d}\right|}{b_d - a_d}\right) = 0. \quad (11.144)$$

Therefore it holds

$$E\left(\frac{T_1 n\left(x_1 - x_{k_1 1}\right)}{b_1 - a_1}, ..., \frac{T_i n\left(x_{(k_i + j_i) i} - x_{k_i i}\right)}{b_i - a_i}, ..., \frac{T_d n\left(x_d - x_{k_d d}\right)}{b_d - a_d}\right) = 0, \quad (11.145)$$

for any arbitrary $(x_1, ..., x_{i-1}, x_{i+1}, ..., x_d) \in \mathbb{R}^{d-1}$.

Let now $x_i = x_{k_i i}$, for all $i = 1, ..., d$, for some $(k_1, ..., k_d) \in \{0, 1, ..., n\}^d$.

Then

$$M_n\left(f, x_{k_1 1}, ..., x_{k_d d}\right) = \frac{f\left(x_{k_1 1}, ..., x_{k_d d}\right) E^*}{E^*} = f\left(x_{k_1 1}, ..., x_{k_d d}\right), \quad (11.146)$$

proving the interpolation property of operators M_n.

Theorem 11.10. *Operators M_n possess the interpolation property over $x_{k_i i}$, $i = 1, ..., d$, $k_i = 0, 1, ..., n$.*

Definition 11.7. Let $f \in C\left(\prod_{i=1}^{d}[a_i, b_i]\right)$. We call

$$\omega_1(f, h) := \sup_{\substack{\text{all } x,y \in \prod_{i=1}^{d}[a_i,b_i]: \\ \|x-y\|_\infty \leq h,}} |f(x) - f(y)| \tag{11.147}$$

$h > 0$, the first multivariate modulus of continuity of f, above $\|\cdot\|_\infty$ is the max-norm.

Approximation result follows

Theorem 11.11. *For $f \in C\left(\prod_{i=1}^{d}[a_i, b_i]\right)$ we have*

$$\|M_n(f) - f\|_\infty \leq \frac{2^d E^*}{E\left(\frac{T_1}{2}, ..., \frac{T_d}{2}\right)} \omega_1\left(f, \frac{\|b-a\|_\infty}{n}\right) =: \varphi_1(n), \tag{11.148}$$

where $\|b - a\|_\infty := \max_{i=1,...,d} \{b_i - a_i\}$.

Proof. Let $x \in \prod_{i=1}^{d}[a_i, b_i]$, we can write

$$M_n(f, x) - f(x)$$

$$= \frac{\sum_{k_1=0}^{n} ... \sum_{k_d=0}^{n} f(x_{k_1 1}, ..., x_{k_d d}) E\left(\frac{T_1 n(x_1 - x_{k_1 1})}{b_1 - a_1}, ..., \frac{T_d n(x_d - x_{k_d d})}{b_d - a_d}\right)}{W} - \frac{f(x) W}{W} \tag{11.149}$$

$$= \frac{\sum_{k_1=0}^{n} ... \sum_{k_d=0}^{n} (f(x_{k_1 1}, ..., x_{k_d d}) - f(x_1, ..., x_d)) E\left(\frac{T_1 n(x_1 - x_{k_1 1})}{b_1 - a_1}, ..., \frac{T_d n(x_d - x_{k_d d})}{b_d - a_d}\right)}{W}. \tag{11.150}$$

Therefore

$$|M_n(f, x) - f(x)| \leq \frac{1}{E\left(\frac{T_1}{2}, ..., \frac{T_d}{2}\right)}$$

$$\cdot \left\{\sum_{k_1=0}^{n} ... \sum_{k_d=0}^{n} |f(x_{k_1 1}, ..., x_{k_d d}) - f(x_1, ..., x_d)| \right. \tag{11.151}$$

$$\left. \cdot E\left(\frac{T_1 n(x_1 - x_{k_1 1})}{b_1 - a_1}, ..., \frac{T_d n(x_d - x_{k_d d})}{b_d - a_d}\right)\right\}$$

$$\leq \frac{1}{E\left(\frac{T_1}{2}, ..., \frac{T_d}{2}\right)} \{0$$

$$+ \sum_{\left\{\substack{\text{all } (k_1,...,k_d) \in \{0,1,...,n\}^d \\ |x_i - x_{k_i i}| < h_i,\ i=1,...,d}\right\}} |f(x_{k_1 1}, ..., x_{k_d d}) - f(x_1, ..., x_d)| \tag{11.152}$$

$$\cdot E\left(\frac{T_1 n\left(x_1 - x_{k_1 1}\right)}{b_1 - a_1}, ..., \frac{T_d n\left(x_d - x_{k_d d}\right)}{b_d - a_d}\right)\right\} \leq$$

(indeed x belongs to a specific box $\prod_{i=1}^{d}\left[x_{k_i i}, x_{(k_i+1)i}\right]$)

$$\frac{2^d E^*}{E\left(\frac{T_1}{2}, ..., \frac{T_d}{2}\right)} \omega_1\left(f, \frac{\|b-a\|_\infty}{n}\right), \tag{11.153}$$

proving the claim. □

Next we denote by $f_{\widetilde{\alpha}} := \frac{\partial^{\widetilde{\alpha}} f}{\partial x^{\widetilde{\alpha}}}$, where $\widetilde{\alpha} := (\alpha_1, ..., \alpha_d)$, $\alpha_i \in \mathbb{Z}^+$, $i = 1, ..., d$, such that $|\widetilde{\alpha}| := \sum_{i=1}^{d} \alpha_i = j$, $j = 1, ..., N$.

High speed approximation using smoothness follows.

Theorem 11.12. *Let $f \in C^N\left(\prod_{i=1}^{d}[a_i, b_i]\right)$, $N \in \mathbb{N}$, and $x \in \prod_{i=1}^{d}[a_i, b_i]$. Then*

i)

$$\left| M_n(f,x) - f(x) - \sum_{j=1}^{N}\left(\sum_{|\widetilde{\alpha}|=j}\left(\frac{f_{\widetilde{\alpha}}(x)}{\prod_{i=1}^{d}\alpha_i!}\right) M_n\left(\prod_{i=1}^{d}(\cdot - x_i)^{\alpha_i}, x\right)\right)\right|$$

$$\leq \frac{2^d E^*}{E\left(\frac{T_1}{2}, ..., \frac{T_d}{2}\right)} \frac{\|b-a\|_\infty^N d^N}{N! n^N} \max_{\widetilde{\alpha}:|\widetilde{\alpha}|=N} \omega_1\left(f_{\widetilde{\alpha}}, \frac{\|b-a\|_\infty}{n}\right), \tag{11.154}$$

ii) assume more that $f_{\widetilde{\alpha}}(x) = 0$, for all $\widetilde{\alpha} : |\widetilde{\alpha}| = 1, ..., N$; where $x \in \prod_{i=1}^{d}[a_i, b_i]$ is fixed, we obtain

$$|M_n(f,x) - f(x)|$$
$$\leq \frac{2^d E^*}{E\left(\frac{T_1}{2}, ..., \frac{T_d}{2}\right)} \frac{\|b-a\|_\infty^N d^N}{N! n^N} \max_{\widetilde{\alpha}:|\widetilde{\alpha}|=N} \omega_1\left(f_{\widetilde{\alpha}}, \frac{\|b-a\|_\infty}{n}\right), \tag{11.155}$$

with high speed of pointwise convergence at $\frac{1}{n^{N+1}}$,

iii)

$$|M_n(f,x) - f(x)| \leq \frac{2^d E^*}{E\left(\frac{T_1}{2}, ..., \frac{T_d}{2}\right)} \cdot$$

$$\left[\sum_{j=1}^{N}\left(\frac{\|b-a\|_\infty^j}{n^j}\right)\left(\sum_{|\widetilde{\alpha}|=j}\left(\frac{|f_{\widetilde{\alpha}}(x)|}{\prod_{i=1}^{d}\alpha_i!}\right)\right)\right. \tag{11.156}$$

$$+\frac{\|b-a\|_{\infty}^{N} d^{N}}{N! n^{N}} \max_{\widetilde{\alpha}:|\widetilde{\alpha}|=N} \omega_1\left(f_{\widetilde{\alpha}}, \frac{\|b-a\|_{\infty}}{n}\right)\Bigg],$$

iv)

$$\|M_n(f)-f\|_{\infty} \leq \frac{2^d E^*}{E\left(\frac{T_1}{2},\ldots,\frac{T_d}{2}\right)} \cdot$$

$$\left[\sum_{j=1}^{N}\left(\frac{\|b-a\|_{\infty}^{j}}{n^{j}}\right)\left(\sum_{|\widetilde{\alpha}|=j}\left(\frac{\|f_{\widetilde{\alpha}}\|_{\infty}}{\prod_{i=1}^{d}\alpha_i!}\right)\right)\right. \tag{11.157}$$

$$\left.+\frac{\|b-a\|_{\infty}^{N} d^{N}}{N! n^{N}} \max_{\widetilde{\alpha}:|\widetilde{\alpha}|=N} \omega_1\left(f_{\widetilde{\alpha}}, \frac{\|b-a\|_{\infty}}{n}\right)\right] =: \varphi_2(n).$$

Proof. Here $f \in C^N\left(\prod_{i=1}^{d}[a_i,b_i]\right)$, $N \in \mathbb{N}$. We call $x_k = (x_{k_1 1},\ldots,x_{k_d d})$. Set

$$g_{x_k}(t) := f(x+t(x_k-x)), \quad 0 \leq t \leq 1, \tag{11.158}$$

$x \in \prod_{i=1}^{d}[a_i,b_i]$, $x=(x_1,\ldots,x_d)$. Then

$$g_{x_k}^{(j)}(t) = \left[\left(\sum_{i=1}^{d}(x_{k_i i}-x_i)\frac{\partial}{\partial x_i}\right)^{j} f\right](x_1+t(x_{k_1 1}-x_1),\ldots,x_d+t(x_{k_d d}-x_d)), \tag{11.159}$$

$$g_{x_k}^{(j)}(0) = \left[\left(\sum_{i=1}^{d}(x_{k_i i}-x_i)\frac{\partial}{\partial x_i}\right)^{j} f\right](x), \tag{11.160}$$

and

$$g_{x_k}(0) = f(x).$$

By Taylor's formula, we get

$$f(x_{k_1 1},\ldots,x_{k_d d}) = g_{x_k}(1) = \sum_{j=0}^{N}\frac{g_{x_k}^{(j)}(0)}{j!} + R_N(x_k,0), \tag{11.161}$$

where

$$R_N(x_k,0) := \int_0^1\left(\int_0^{t_1}\ldots\left(\int_0^{t_{N-1}}\left(g_{x_k}^{(N)}(t_N)-g_{x_k}^{(N)}(0)\right)dt_N\right)\ldots\right)dt_1. \tag{11.162}$$

Thus,

$$\frac{f(x_{k_1 1},\ldots,x_{k_d d})\, E\left(\frac{T_1 n\left(x_1-x_{k_1 1}\right)}{b_1-a_1},\ldots,\frac{T_d n\left(x_d-x_{k_d d}\right)}{b_d-a_d}\right)}{W} \tag{11.163}$$

$$= \sum_{j=0}^{N} \frac{g_{x_k}^{(j)}(0)}{j!} \frac{E\left(\frac{T_1 n\left(x_1 - x_{k_1 1}\right)}{b_1 - a_1}, ..., \frac{T_d n\left(x_d - x_{k_d d}\right)}{b_d - a_d}\right)}{W}$$

$$+ \frac{E\left(\frac{T_1 n\left(x_1 - x_{k_1 1}\right)}{b_1 - a_1}, ..., \frac{T_d n\left(x_d - x_{k_d d}\right)}{b_d - a_d}\right)}{W} R_N(x_k, 0). \tag{11.164}$$

Therefore

$$M_n(f, x) - f(x)$$

$$= \sum_{j=1}^{N} \frac{1}{j!} \left(\frac{\sum_{k_1=0}^{n} \cdots \sum_{k_d=0}^{n} g_{x_k}^{(j)}(0) E\left(\frac{T_1 n\left(x_1 - x_{k_1 1}\right)}{b_1 - a_1}, ..., \frac{T_d n\left(x_d - x_{k_d d}\right)}{b_d - a_d}\right)}{W} \right) + R^*, \tag{11.165}$$

where

$$R^* := \frac{\sum_{k_1=0}^{n} \cdots \sum_{k_d=0}^{n} E\left(\frac{T_1 n\left(x_1 - x_{k_1 1}\right)}{b_1 - a_1}, ..., \frac{T_d n\left(x_d - x_{k_d d}\right)}{b_d - a_d}\right)}{W} R_N(x_k, 0). \tag{11.166}$$

Consequently, we obtain

$$|M_n(f, x) - f(x)| \leq$$

$$\sum_{j=1}^{N} \frac{1}{j!} \frac{\sum_{k_1=0}^{n} \cdots \sum_{k_d=0}^{n} \left|g_{x_k}^{(j)}(0)\right| E\left(\frac{T_1 n\left(x_1 - x_{k_1 1}\right)}{b_1 - a_1}, ..., \frac{T_d n\left(x_d - x_{k_d d}\right)}{b_d - a_d}\right)}{E\left(\frac{T_1}{2}, ..., \frac{T_d}{2}\right)} + |R^*|$$

$$\leq \sum_{j=1}^{N} \frac{1}{j!} \frac{2^d \left(\frac{\|b-a\|_\infty^j}{n^j}\right) \left(\left(\sum_{i=1}^{d} \left|\frac{\partial}{\partial x_i}\right|\right)^j f(x)\right) E^*}{E\left(\frac{T_1}{2}, ..., \frac{T_d}{2}\right)} + |R^*| \tag{11.167}$$

$$= \frac{2^d E^*}{E\left(\frac{T_1}{2}, ..., \frac{T_d}{2}\right)} \left[\sum_{j=1}^{N} \frac{1}{j!} \left(\left(\sum_{i=1}^{d} \left|\frac{\partial}{\partial x_i}\right|\right)^j f(x)\right) \left(\frac{\|b-a\|_\infty^j}{n^j}\right)\right] + |R^*|. \tag{11.168}$$

Next, we estimate $|R^*|$.

For that, we observe

$$|R^*| \leq \frac{\sum_{k_1=0}^{n} \cdots \sum_{k_d=0}^{n} E\left(\frac{T_1 n\left(x_1 - x_{k_1 1}\right)}{b_1 - a_1}, ..., \frac{T_d n\left(x_d - x_{k_d d}\right)}{b_d - a_d}\right)}{E\left(\frac{T_1}{2}, ..., \frac{T_d}{2}\right)}. \tag{11.169}$$

$$\left(\int_0^1 \left(\int_0^{t_1} \cdots \left(\int_0^{t_{N-1}} \left|g_{x_k}^{(N)}(t_N) - g_{x_k}^{(N)}(0)\right| dt_N\right) \cdots\right) dt_1\right)$$

$$= \frac{\sum_{k_1=0}^{n} \dots \sum_{k_d=0}^{n} E\left(\frac{T_1 n(x_1 - x_{k_1 1})}{b_1 - a_1}, \dots, \frac{T_d n(x_d - x_{k_d d})}{b_d - a_d}\right)}{E\left(\frac{T_1}{2}, \dots, \frac{T_d}{2}\right)}.$$

$$\left(\int_0^1 \left(\int_0^{t_1} \dots \left(\int_0^{t_{N-1}} \left| \left[\left(\sum_{i=1}^{d} (x_{k_i i} - x_i) \frac{\partial}{\partial x_i}\right)^N f\right]\right.\right.\right.\right.$$

$$(x_1 + t_N (x_{k_1 1} - x_1), \dots, x_d + t_N (x_{k_d d} - x_d))$$

$$\left.\left.\left.\left. - \left[\left(\sum_{i=1}^{d} (x_{k_i i} - x_i) \frac{\partial}{\partial x_i}\right)^N f\right] (x_1, \dots, x_d) \right| dt_N\right) \dots\right) dt_1\right) \quad (11.170)$$

$$\leq \frac{2^d E^*}{E\left(\frac{T_1}{2}, \dots, \frac{T_d}{2}\right)} \left(\int_0^1 \left(\int_0^{t_1} \dots \left(\int_0^{t_{N-1}} \left\{\left(\frac{\|b - a\|_\infty^N}{n^N}\right) d^N\right.\right.\right.\right.$$

$$\left.\left.\left.\left.\max_{\widetilde{\alpha}:|\alpha|=N} \omega_1\left(f_{\widetilde{\alpha}}, \frac{\|b - a\|_\infty}{n}\right)\right\} dt_N\right) \dots\right) dt_1\right)$$

$$= \frac{2^d E^*}{N! E\left(\frac{T_1}{2}, \dots, \frac{T_d}{2}\right)} \frac{\|b - a\|_\infty^N d^N}{n^N} \max_{\widetilde{\alpha}:|\alpha|=N} \omega_1\left(f_{\widetilde{\alpha}}, \frac{\|b - a\|_\infty}{n}\right). \quad (11.171)$$

That is

$$|R^*| \leq \frac{2^d E^*}{N! E\left(\frac{T_1}{2}, \dots, \frac{T_d}{2}\right)} \frac{\|b - a\|_\infty^N d^N}{n^N} \max_{\widetilde{\alpha}:|\alpha|=N} \omega_1\left(f_{\widetilde{\alpha}}, \frac{\|b - a\|_\infty}{n}\right). \quad (11.172)$$

The proof of the theorem now is complete. □

About multivariate Taylor formula and estimates (see [15], pp. 284-286)

Let $\prod_{i=1}^{d} [a_i, b_i]$; $d \geq 2$; $z := (z_1, \dots, z_d)$, $x_0 := (x_{01}, \dots, x_{0d}) \in \prod_{i=1}^{d} [a_i, b_i]$. We consider the space of functions $AC^N\left(\prod_{i=1}^{d} [a_i, b_i]\right)$ with $f : \prod_{i=1}^{d} [a_i, b_i] \to \mathbb{R}$ be such that all partial derivatives of order $(N - 1)$ are coordinate-wise absolutely continuous functions on $\prod_{i=1}^{d} [a_i, b_i]$, $N \in \mathbb{N}$. Also $f \in C^{N-1}\left(\prod_{i=1}^{d} [a_i, b_i]\right)$. Each Nth order partial derivative is denoted by $f_{\widetilde{\alpha}} := \frac{\partial^{\widetilde{\alpha}} f}{\partial x^{\widetilde{\alpha}}}$, where $\widetilde{\alpha} := (\alpha_1, \dots, \alpha_d)$, $\alpha_i \in \mathbb{Z}^+$, $i = 1, \dots, d$ and $|\widetilde{\alpha}| := \sum_{i=1}^{d} \alpha_i = N$. Consider $g_z(t) := f(x_0 + t(z - x_0))$, $t \geq 0$. Then

$$g_z^{(j)}(t) = \left[\left(\sum_{i=1}^{d} (z_i - x_{0i}) \frac{\partial}{\partial x_i}\right)^j f\right] (x_{01} + t(z_1 - x_{01}), \dots, x_{0d} + t(z_N - x_{0d})), \quad (11.173)$$

for all $j = 0, 1, 2, ..., N$.

We mention the following multivariate Taylor theorem.

Theorem 11.13. *Under the above assumptions we have*

$$f(z_1, ..., z_d) = g_z(1) = \sum_{j=0}^{N-1} \frac{g_z^{(j)}(0)}{j!} + R_N(z, 0), \tag{11.174}$$

where

$$R_N(z, 0) := \int_0^1 \left(\int_0^{t_1} ... \left(\int_0^{t_{N-1}} g_z^{(N)}(t_N) \, dt_N \right) ... \right) dt_1, \tag{11.175}$$

or

$$R_N(z, 0) = \frac{1}{(N-1)!} \int_0^1 (1-\theta)^{N-1} g_z^{(N)}(\theta) \, d\theta. \tag{11.176}$$

Notice that $g_z(0) = f(x_0)$.

We make

Remark 11.9. Assume here that

$$\|f_{\widetilde{\alpha}}\|_{\infty,N}^{\max} := \max_{|\widetilde{\alpha}|=N} \|f_{\widetilde{\alpha}}\|_{\infty} < \infty. \tag{11.177}$$

Then

$$\left\| g_z^{(N)} \right\|_{\infty,[0,1]} = \left\| \left[\left(\sum_{i=1}^{d} (z_i - x_{0i}) \frac{\partial}{\partial x_i} \right)^N f \right] (x_0 + t(z - x_0)) \right\|_{\infty,[0,1]}$$

$$\leq \left(\sum_{i=1}^{d} |z_i - x_{0i}| \right)^N \|f_{\widetilde{\alpha}}\|_{\infty,N}^{\max}, \tag{11.178}$$

that is

$$\left\| g_z^{(N)} \right\|_{\infty,[0,1]} \leq \left(\|z - x_0\|_{l_1} \right)^N \|f_{\widetilde{\alpha}}\|_{\infty,N}^{\max} < \infty. \tag{11.179}$$

Hence we get by (11.176) that

$$|R_N(z, 0)| \leq \frac{\left\| g_z^{(N)} \right\|_{\infty,[0,1]}}{N!} < \infty. \tag{11.180}$$

And it holds

$$|R_N(z, 0)| \leq \frac{\left(\|z - x_0\|_{l_1} \right)^N}{N!} \|f_{\widetilde{\alpha}}\|_{\infty,N}^{\max}, \tag{11.181}$$

$\forall \, z, x_0 \in \prod_{i=1}^{d} [a_i, b_i]$.

We will use decisively (11.181).

Next follows a multivariate Voronovskaya type asymptotic expansion

Theorem 11.14. *Let* $f \in AC^N\left(\prod_{i=1}^{d}[a_i,b_i]\right)$, $d \in \mathbb{N}-\{1\}$, $N \in \mathbb{N}$, *with*

$$\|f_{\widetilde{\alpha}}\|_{\infty,N}^{\max} := \max_{|\widetilde{\alpha}|=N} \|f_{\widetilde{\alpha}}\|_{\infty} < \infty. \tag{11.182}$$

Then

$$M_n(f,x) - f(x) - \sum_{j=1}^{N-1}\left(\sum_{\substack{\widetilde{\alpha}:=(\alpha_1,\ldots,\alpha_d),\ \alpha_i\in\mathbb{Z}^+\\ i=1,\ldots,d,\ |\widetilde{\alpha}|:=\sum_{i=1}^{d}\alpha_i=j}}\left(\frac{f_{\widetilde{\alpha}}(x)}{\prod_{i=1}^{d}\alpha_i!}\right)M_n\left(\prod_{i=1}^{d}(\cdot-x_i)^{\alpha_i},x\right)\right)$$

$$= o\left(\frac{1}{n^{N-\varepsilon}}\right), \quad 0<\varepsilon\leq N. \tag{11.183}$$

If $N=1$, *the sum collapses.*

The last (11.183) implies

$$n^{N-\varepsilon}\left[M_n(f,x) - f(x) - \right.$$

$$\left.\sum_{j=1}^{N-1}\left(\sum_{\substack{\widetilde{\alpha}:=(\alpha_1,\ldots,\alpha_d),\ \alpha_i\in\mathbb{Z}^+\\ i=1,\ldots,d,\ |\widetilde{\alpha}|:=\sum_{i=1}^{d}\alpha_i=j}}\left(\frac{f_{\widetilde{\alpha}}(x)}{\prod_{i=1}^{d}\alpha_i!}\right)M_n\left(\prod_{i=1}^{d}(\cdot-x_i)^{\alpha_i},x\right)\right)\right]\to 0, \tag{11.184}$$

as $n\to\infty$, $0<\varepsilon\leq N$.

When $N=1$ *or* $f_{\widetilde{\alpha}}(x)=0$, *all* $\widetilde{\alpha}:|\widetilde{\alpha}|=j=1,\ldots,N-1$, *then*

$$n^{N-\varepsilon}\left[(M_n(f))(x)-f(x)\right]\to 0, \tag{11.185}$$

as $n\to\infty$, $0<\varepsilon\leq N$.

Proof. We call $x_k=(x_{k_1 1},\ldots,x_{k_d d})$. Set

$$g_{x_k}(t) := f(x+t(x_k-x)), \quad 0\leq t\leq 1, \tag{11.186}$$

$x\in\prod_{i=1}^{d}[a_i,b_i]$. Then

$$g_{x_k}^{(j)}(t) = \left[\left(\sum_{i=1}^{d}(x_{k_i i}-x_i)\frac{\partial}{\partial x_i}\right)^j f\right](x_1+t(x_{k_1 1}-x_1),\ldots,x_d+t(x_{k_d d}-x_d)), \tag{11.187}$$

and

$$g_{x_k}(0) = f(x).$$

By Taylor's formula, we get

$$f(x_k) = g_{x_k}(1) = \sum_{j=0}^{N-1} \frac{g_{x_k}^{(j)}(0)}{j!} + R_N(x_k, 0), \tag{11.188}$$

where

$$R_N(x_k, 0) := \frac{1}{(N-1)!} \int_0^1 (1-\theta)^{N-1} g_{x_k}^{(N)}(\theta)\, d\theta. \tag{11.189}$$

Here we denote by $f_{\widetilde{\alpha}} := \frac{\partial^{\widetilde{\alpha}} f}{\partial x^{\widetilde{\alpha}}}$, $\widetilde{\alpha} := (\alpha_1, ..., \alpha_d)$, $\alpha_i \in \mathbb{Z}^+$, $i = 1, ..., d$, such that $|\widetilde{\alpha}| := \sum_{i=1}^{d} \alpha_i = N$. Thus

$$\frac{f(x_k) E\left(\frac{T_1 n(x_1 - x_{k_1 1})}{b_1 - a_1}, ..., \frac{T_d n(x_d - x_{k_d d})}{b_d - a_d}\right)}{W} \tag{11.190}$$

$$= \sum_{j=0}^{N-1} \frac{g_{x_k}^{(j)}(0)}{j!} \frac{E\left(\frac{T_1 n(x_1 - x_{k_1 1})}{b_1 - a_1}, ..., \frac{T_d n(x_d - x_{k_d d})}{b_d - a_d}\right)}{W}$$

$$+ \frac{E\left(\frac{T_1 n(x_1 - x_{k_1 1})}{b_1 - a_1}, ..., \frac{T_d n(x_d - x_{k_d d})}{b_d - a_d}\right)}{W} R_N(x_k, 0). \tag{11.191}$$

Therefore it holds

$$M_n(f, x) - f(x)$$

$$- \sum_{j=1}^{N-1} \frac{1}{j!} \frac{\left(\sum_{k_1=0}^{n} \cdots \sum_{k_d=0}^{n} g_{x_k}^{(j)}(0) E\left(\frac{T_1 n(x_1 - x_{k_1 1})}{b_1 - a_1}, ..., \frac{T_d n(x_d - x_{k_d d})}{b_d - a_d}\right)\right)}{W} = R^*, \tag{11.192}$$

where

$$R^* := \frac{\sum_{k_1=0}^{n} \cdots \sum_{k_d=0}^{n} E\left(\frac{T_1 n(x_1 - x_{k_1 1})}{b_1 - a_1}, ..., \frac{T_d n(x_d - x_{k_d d})}{b_d - a_d}\right)}{W} R_N(x_k, 0). \tag{11.193}$$

Hence

$$M_n(f, x) - f(x) - \sum_{j=1}^{N-1} \left(\sum_{\substack{\widetilde{\alpha} := (\alpha_1, ..., \alpha_d),\ \alpha_i \in \mathbb{Z}^+ \\ i=1,...,d,\ |\widetilde{\alpha}| := \sum_{i=1}^{d} \alpha_i = j}} \left(\frac{f_{\widetilde{\alpha}}(x)}{\prod_{i=1}^{d} \alpha_i!} \right) M_n\left(\prod_{i=1}^{d} (\cdot - x_i)^{\alpha_i}, x \right) \right)$$

$$= R^*. \tag{11.194}$$

Notice that

$$R^* = \frac{\sum_{k_1=0}^{n} \cdots \sum_{k_d=0}^{n} E\left(\frac{T_1 n\left(x_1 - x_{k_1 1}\right)}{b_1 - a_1}, \ldots, \frac{T_d n\left(x_d - x_{k_d d}\right)}{b_d - a_d}\right)}{W}$$

$$N \int_0^1 (1-\theta)^{N-1} \sum_{\substack{\widetilde{\alpha} := (\alpha_1, \ldots, \alpha_d),\ \alpha_i \in \mathbb{Z}^+ \\ i=1,\ldots,d,\ |\widetilde{\alpha}| := \sum_{i=1}^{d} \alpha_i = N}} \left(\frac{1}{\prod_{i=1}^{d} \alpha_i!} \right)$$

$$\left(\prod_{i=1}^{d} \left(x_{k_i i} - x_i\right)^{\alpha_i} \right) f_{\widetilde{\alpha}} \left(x + \theta \left(x_k - x\right)\right) d\theta. \tag{11.195}$$

Hence it holds

$$|R^*| \overset{(11.181)}{\leq} \frac{\sum_{k_1=0}^{n} \cdots \sum_{k_d=0}^{n} E\left(\frac{T_1 n\left(x_1 - x_{k_1 1}\right)}{b_1 - a_1}, \ldots, \frac{T_d n\left(x_d - x_{k_d d}\right)}{b_d - a_d}\right)}{E\left(\frac{T_1}{2}, \ldots, \frac{T_d}{2}\right)} \tag{11.196}$$

$$\frac{\left(\|x_k - x\|_{l_1}\right)^N}{N!} \|f_{\widetilde{\alpha}}\|_{\infty,N}^{\max} \leq$$

$$\frac{2^d E^*}{E\left(\frac{T_1}{2}, \ldots, \frac{T_d}{2}\right)} \left(d \frac{\|b-a\|_\infty}{n} \right)^N \frac{\|f_{\widetilde{\alpha}}\|_{\infty,N}^{\max}}{N!}. \tag{11.197}$$

That is

$$|R^*| \leq \frac{\delta}{n^N}, \tag{11.198}$$

where

$$\delta := \frac{2^d E^* d^N \|b-a\|_\infty^N \|f_{\widetilde{\alpha}}\|_{\infty,N}^{\max}}{E\left(\frac{T_1}{2}, \ldots, \frac{T_d}{2}\right) N!} < +\infty. \tag{11.199}$$

That is

$$|R^*| = O\left(\frac{1}{n^N}\right), \tag{11.200}$$

and

$$|R^*| = o(1). \tag{11.201}$$

And letting $0 < \varepsilon \leq N$, we derive

$$\frac{|R^*|}{\left(\frac{1}{n^{N-\varepsilon}}\right)} \leq \frac{\delta}{n^\varepsilon} \to 0, \tag{11.202}$$

as $n \to \infty$.

That is

$$|R^*| = o\left(\frac{1}{n^{N-\varepsilon}}\right). \tag{11.203}$$

The proof is completed. ∎

11.2.3 *Neural Networks Iterated Approximation and Interpolation*

We make

Remark 11.10. Here E is assumed additionally to be continuous.

Let $f \in C\left(\prod_{i=1}^{d}[a_i,b_i]\right)$. We (see (11.138), (11.140)) proved that $W>0$. Hence $M_n(f) \in C\left(\prod_{i=1}^{d}[a_i,b_i]\right)$. Furthermore $M_n(f)-f \in C\left(\prod_{i=1}^{d}[a_i,b_i]\right)$.

We proved earlier (11.133) that

$$\|M_n(f)\|_\infty \leq \|f\|_\infty < +\infty. \tag{11.204}$$

Clearly then

$$\left\|M_n^2(f)\right\|_\infty = \|M_n(M_n(f))\|_\infty \leq \|M_n(f)\|_\infty \leq \|f\|_\infty . \tag{11.205}$$

Therefore

$$\left\|M_n^k(f)\right\|_\infty \leq \|f\|_\infty, \quad \forall\, k \in \mathbb{N}. \tag{11.206}$$

Also we see that

$$\left\|M_n^k(f)\right\|_\infty \leq \left\|M_n^{k-1}(f)\right\|_\infty \leq \dots \leq \|M_n(f)\|_\infty \leq \|f\|_\infty . \tag{11.207}$$

Also it holds

$$M_n(1)=1, \;\; M_n^k(1)=1, \;\; \forall\, k \in \mathbb{N}. \tag{11.208}$$

Here M_n^k are positive linear operators.

Call $x_k=(x_{k_1 1},\dots,x_{k_d d})$, we proved (11.146), that

$$(M_n(f))(x_k)=f(x_k), \tag{11.209}$$

the interpolation property of M_n.

Hence we get

$$\left(M_n^2(f)\right)(x_k)=(M_n(M_n(f)))(x_k)$$

(by Theorem 11.10)

$$=(M_n(f))(x_k)=f(x_k), \tag{11.210}$$

In general it holds

$$\left(M_n^k(f)\right)(x_k)=f(x_k), \;\; \forall\, k \in \mathbb{N}, \tag{11.211}$$

proving interpolation of the operators M_n^k.

Remark 11.11. Let $r \in \mathbb{N}$ and M_n as above. We observe that

$$\begin{aligned} M_n^r f - f &= \left(M_n^r f - M_n^{r-1}f\right) + \left(M_n^{r-1}f - M_n^{r-2}f\right) \\ &+ \left(M_n^{r-2}f - M_n^{r-3}f\right) + \dots + \left(M_n^2 f - M_n f\right) + (M_n f - f). \end{aligned} \tag{11.212}$$

Then

$$\begin{aligned} \|M_n^r f - f\|_\infty &\leq \left\|M_n^r f - M_n^{r-1}f\right\|_\infty + \left\|M_n^{r-1}f - M_n^{r-2}f\right\|_\infty \\ &+ \left\|M_n^{r-2}f - M_n^{r-3}f\right\|_\infty + \dots + \left\|M_n^2 f - M_n f\right\|_\infty + \|M_n f - f\|_\infty \\ &= \left\|M_n^{r-1}(M_n f - f)\right\|_\infty + \left\|M_n^{r-2}(M_n f - f)\right\|_\infty + \left\|M_n^{r-3}(M_n f - f)\right\|_\infty \\ &+ \dots + \|M_n(M_n f - f)\|_\infty + \|M_n f - f\|_\infty \leq r\,\|M_n f - f\|_\infty . \end{aligned} \tag{11.213}$$

That is

$$\|M_n^r f - f\|_\infty \leq r\,\|M_n f - f\|_\infty . \tag{11.214}$$

Remark 11.12 (Conclusion). Thus, the speed of convergence to the unit operator of M_n^r is not worse than of M_n.

Remark 11.13. Let $m_1, ..., m_r \in \mathbb{N} : m_1 \leq m_2 \leq ... \leq m_r$, $r \in \mathbb{N}$.
Let M_{m_i} as above, $i = 1, ..., r$.

Then it holds

$$\begin{aligned} &M_{m_r}\left(M_{m_{r-1}}\left(...M_{m_2}\left(M_{m_1}(f)\right)\right)\right) - f \\ &= \left[M_{m_r}\left(M_{m_{r-1}}\left(...M_{m_2}\left(M_{m_1}(f)\right)\right)\right) - M_{m_r}\left(M_{m_{r-1}}\left(...M_{m_2}(f)\right)\right)\right] \\ &\quad + \left[M_{m_r}\left(M_{m_{r-1}}\left(...M_{m_3}\left(M_{m_2}(f)\right)\right)\right) - M_{m_r}\left(M_{m_{r-1}}\left(...M_{m_3}(f)\right)\right)\right] \end{aligned} \quad (11.215)$$

$$\begin{aligned} &\quad + \left[M_{m_r}\left(M_{m_{r-1}}\left(...M_{m_4}\left(M_{m_3}(f)\right)\right)\right) - M_{m_r}\left(M_{m_{r-1}}\left(...M_{m_4}(f)\right)\right)\right] \\ &\quad + ... + \left[M_{m_r}\left(M_{m_{r-1}}f\right) - M_{m_r}f\right] + \left[M_{m_r}f - f\right] \\ &= \left[M_{m_r}\left(M_{m_{r-1}}\left(...M_{m_2}\right)\right)\left(M_{m_1}f - f\right)\right] \\ &\quad + \left[M_{m_r}\left(M_{m_{r-1}}\left(...M_{m_3}\right)\right)\left(M_{m_2}f - f\right)\right] \\ &\quad + \left[M_{m_r}\left(M_{m_{r-1}}\left(...M_{m_4}\right)\right)\left(M_{m_3}f - f\right)\right] \\ &\quad + ... + \left[M_{m_r}\left(M_{m_{r-1}}f - f\right)\right] + \left[M_{m_r}f - f\right]. \end{aligned} \quad (11.216)$$

Therefore

$$\begin{aligned} &\left\|M_{m_r}\left(M_{m_{r-1}}\left(...M_{m_2}\left(M_{m_1}(f)\right)\right)\right) - f\right\|_\infty \\ &\leq \left\|M_{m_r}\left(M_{m_{r-1}}\left(...M_{m_2}\right)\right)\left(M_{m_1}f - f\right)\right\|_\infty \quad (11.217) \\ &\quad + \left\|M_{m_r}\left(M_{m_{r-1}}\left(...M_{m_3}\right)\right)\left(M_{m_2}f - f\right)\right\|_\infty \\ &\quad + \left\|M_{m_r}\left(M_{m_{r-1}}\left(...M_{m_4}\right)\right)\left(M_{m_3}f - f\right)\right\|_\infty \\ &\quad + ... + \left\|M_{m_r}\left(M_{m_{r-1}}f - f\right)\right\|_\infty + \left\|M_{m_r}f - f\right\|_\infty \\ &\leq \left\|M_{m_1}f - f\right\|_\infty + \left\|M_{m_2}f - f\right\|_\infty + \left\|M_{m_3}f - f\right\|_\infty \quad (11.218) \\ &\quad + ... + \left\|M_{m_{r-1}}f - f\right\|_\infty + \left\|M_{m_r}f - f\right\|_\infty = \sum_{i=1}^{r} \left\|M_{m_i}f - f\right\|_\infty. \end{aligned} \quad (11.219)$$

We have proved that

$$\left\|M_{m_r}\left(M_{m_{r-1}}\left(...M_{m_2}\left(M_{m_1}(f)\right)\right)\right) - f\right\|_\infty \leq \sum_{i=1}^{r} \left\|M_{m_i}f - f\right\|_\infty. \quad (11.220)$$

Using (11.214) we derive

Theorem 11.15. *Let* $f \in C\left(\prod_{i=1}^{d} [a_i, b_i]\right)$, $r \in \mathbb{N}$. *Then*

$$\left\|M_n^r f - f\right\|_\infty \leq \frac{r 2^d E^*}{E\left(\frac{T_1}{2}, ..., \frac{T_d}{2}\right)} \omega_1\left(f, \frac{\|b - a\|_\infty}{n}\right). \quad (11.221)$$

Proof. Also use of (11.148). □

Theorem 11.16. *Let* $f \in C^N\left(\prod_{i=1}^{d}[a_i, b_i]\right)$, $N \in \mathbb{N}$, $r \in \mathbb{N}$. *Then*

$$\|M_n^r f - f\|_\infty \leq r\varphi_2(n), \tag{11.222}$$

where $\varphi_2(n)$ *is as in (11.157).*

Proof. Use also of (11.157). □

Next we use (11.220).

Theorem 11.17. *Let* $m_1, ..., m_r \in \mathbb{N} : m_1 \leq m_2 \leq ... \leq m_r$, $r \in \mathbb{N}$, $f \in C\left(\prod_{i=1}^{d}[a_i, b_i]\right)$. *Then*

$$\left\|M_{m_r}\left(M_{m_{r-1}}\left(...M_{m_2}\left(M_{m_1}(f)\right)\right)\right) - f\right\|_\infty \leq \sum_{i=1}^{r}\varphi_1(m_i) \tag{11.223}$$

$$\leq \frac{r2^d E^*}{E\left(\frac{T_1}{2}, ..., \frac{T_d}{2}\right)}\omega_1\left(f, \frac{\|b-a\|_\infty}{m_1}\right),$$

where φ_1 *as in (11.148).*

Proof. Use also of (11.148). □

Theorem 11.18. *Let* $m_1, ..., m_r \in \mathbb{N} : m_1 \leq m_2 \leq ... \leq m_r$, $r \in \mathbb{N}$, $f \in C^N\left(\prod_{i=1}^{d}[a_i, b_i]\right)$, $N \in \mathbb{N}$. *Then*

$$\left\|M_{m_r}\left(M_{m_{r-1}}\left(...M_{m_2}\left(M_{m_1}(f)\right)\right)\right) - f\right\|_\infty \leq \sum_{i=1}^{r}\varphi_2(m_i)$$

$$\leq \frac{r2^d E^*}{E\left(\frac{T_1}{2}, ..., \frac{T_d}{2}\right)}\left[\sum_{j=1}^{N}\left(\frac{\|b-a\|_\infty^j}{m_1^j}\right)\left(\sum_{|\widetilde{\alpha}|=j}\frac{\|f_{\widetilde{\alpha}}\|_\infty}{\prod_{i=1}^{d}\alpha_i!}\right)\right.$$

$$\left.+\frac{\|b-a\|_\infty^N d^N}{N! m_1^N}\max_{\widetilde{\alpha}:|\widetilde{\alpha}|=N}\omega_1\left(f_{\widetilde{\alpha}}, \frac{\|b-a\|_\infty}{m_1}\right)\right], \tag{11.224}$$

where φ_2 *as in (11.157).*

Proof. Also use of (11.157). □

11.2.4 *Complex Multivariate Neural Network Approximation and Interpolation*

We make

Remark 11.14. Let $f : \prod_{i=1}^{d} [a_i, b_i] \to \mathbb{C}$ with real and imaginary parts $f_1, f_2 : f = f_1 + if_2$, $i = \sqrt{-i}$. Clearly f is continuous iff f_1 and f_2 are continuous.

Given that $f_1, f_2 \in C^N \left(\prod_{i=1}^{d} [a_i, b_i] \right)$, $N \in \mathbb{N}$, it holds

$$f_{\widetilde{\alpha}}(x) = f_{1,\widetilde{\alpha}}(x) + if_{2,\widetilde{\alpha}}(x), \tag{11.225}$$

where $\widetilde{\alpha}$ indicates a partial derivative of any order and arrangement.

Let $f \in C \left(\prod_{i=1}^{d} [a_i, b_i], \mathbb{C} \right)$ the space of continuous functions $f : \prod_{i=1}^{d} [a_i, b_i] \to \mathbb{C}$. Then $f_1, f_2 \in C \left(\prod_{i=1}^{d} [a_i, b_i] \right)$, and thus both are bounded, implying that f is bounded.

We define

$$M_n^{\mathbb{C}}(f, x) := M_n(f_1, x) + iM_n(f_2, x), \quad \forall\, x \in \prod_{i=1}^{d} [a_i, b_i]. \tag{11.226}$$

We observe that

$$\left| M_n^{\mathbb{C}}(f, x) - f(x) \right| \leq |M_n(f_1, x) - f_1(x)| + |M_n(f_2, x) - f_2(x)|, \tag{11.227}$$

and

$$\left\| M_n^{\mathbb{C}}(f) - f \right\|_\infty \leq \|M_n(f_1) - f_1\|_\infty + \|M_n(f_2) - f_2\|_\infty. \tag{11.228}$$

If f is bounded then f_1, f_2 are also bounded.

For the interpolation property we assume that f is bounded and measurable. Thus f_1, f_2 are measurable.

We have (for any $(k_1, ..., k_d) \in \{0, 1, ..., n\}^d$)

$$M_n^{\mathbb{C}}(f, x_{k_1 1}, ..., x_{k_d d}) = M_n(f_1, x_{k_1 1}, ..., x_{k_d d}) + iM_n(f_2, x_{k_1 1}, ..., x_{k_d d})$$

$$= f_1(x_{k_1 1}, ..., x_{k_d d}) + if_2(x_{k_1 1}, ..., x_{k_d d}) = f(x_{k_1 1}, ..., x_{k_d d}), \tag{11.229}$$

proving interpolation of $M_n^{\mathbb{C}}$.

Theorem 11.19. *Let $f \in C \left(\prod_{i=1}^{d} [a_i, b_i], \mathbb{C} \right)$, such that $f = f_1 + if_2$, $n \in \mathbb{N}$. Then*

$$\left\| M_n^{\mathbb{C}}(f) - f \right\|_\infty \leq \frac{2^d E^*}{E\left(\frac{T_1}{2}, ..., \frac{T_d}{2}\right)} \cdot$$

$$\left[\omega_1\left(f_1, \frac{\|b - a\|_\infty}{n}\right) + \omega_1\left(f_2, \frac{\|b - a\|_\infty}{n}\right) \right]. \tag{11.230}$$

Proof. By Theorem 11.11. □

Theorem 11.20. *Let* $f : \prod_{i=1}^{d} [a_i, b_i] \to \mathbb{C}$, *such that* $f = f_1 + if_2$. *Assume* $f_1, f_2 \in C^N \left(\prod_{i=1}^{d} [a_i, b_i] \right)$, $N \in \mathbb{N}$, $n \in \mathbb{N}$. *Then*

$$\left| M_n^{\mathbb{C}} (f, x) - f (x) \right| \leq \frac{2^d E^*}{E \left(\frac{T_1}{2}, ..., \frac{T_d}{2} \right)} \cdot$$

$$\left[\sum_{j=1}^{N} \frac{1}{j!} \left(\frac{\|b-a\|_{\infty}^{j}}{n^j} \right) \left[\left(\left(\sum_{i=1}^{d} \left| \frac{\partial}{\partial x_i} \right| \right)^j f_1 (x) \right) \right. \right.$$

$$\left. + \left(\left(\sum_{i=1}^{d} \left| \frac{\partial}{\partial x_i} \right| \right)^j f_2 (x) \right) \right] + \frac{\|b-a\|_{\infty}^{N} d^N}{N! n^N} \cdot$$

$$\left[\max_{\widetilde{\alpha}:|\widetilde{\alpha}|=N} \omega_1 \left(f_{1,\widetilde{\alpha}}, \frac{\|b-a\|_{\infty}}{n} \right) + \max_{\widetilde{\alpha}:|\widetilde{\alpha}|=N} \omega_1 \left(f_{2,\widetilde{\alpha}}, \frac{\|b-a\|_{\infty}}{n} \right) \right] \Bigg] \tag{11.231}$$

$$= \frac{2^d E^*}{E \left(\frac{T_1}{2}, ..., \frac{T_d}{2} \right)} \left[\sum_{j=1}^{N} \left(\frac{\|b-a\|_{\infty}^{j}}{n^j} \right) \left(\sum_{|\widetilde{\alpha}|=j} \left(\frac{|f_{1,\widetilde{\alpha}} (x)| + |f_{2,\widetilde{\alpha}} (x)|}{\prod_{i=1}^{d} \alpha_i!} \right) \right) \right.$$

$$\left. + \frac{\|b-a\|_{\infty}^{N} d^N}{N! n^N} \left[\max_{\widetilde{\alpha}:|\widetilde{\alpha}|=N} \omega_1 \left(f_{1,\widetilde{\alpha}}, \frac{\|b-a\|_{\infty}}{n} \right) + \max_{\widetilde{\alpha}:|\widetilde{\alpha}|=N} \omega_1 \left(f_{2,\widetilde{\alpha}}, \frac{\|b-a\|_{\infty}}{n} \right) \right] \right]. \tag{11.232}$$

Proof. By (11.156). □

11.2.5 *Fuzzy Fractional Mathematical Analysis Background*

We need the following basic background

Definition 11.8. (see [42]) Let $\mu : \mathbb{R} \to [0, 1]$ with the following properties:

(i) is normal, i.e., $\exists\ x_0 \in \mathbb{R}$; $\mu (x_0) = 1$.

(ii) $\mu (\lambda x + (1 - \lambda) y) \geq \min\{\mu (x), \mu (y)\}$, $\forall\ x, y \in \mathbb{R}$, $\forall\ \lambda \in [0, 1]$ (μ is called a convex fuzzy subset).

(iii) μ is upper semicontinuous on $\mathbb{R}$, i.e. $\forall\ x_0 \in \mathbb{R}$ and $\forall\ \varepsilon > 0$, $\exists$ neighborhood $V (x_0) : \mu (x) \leq \mu (x_0) + \varepsilon$, $\forall\ x \in V (x_0)$.

(iv) The set $\overline{\text{supp} (\mu)}$ is compact in $\mathbb{R}$ (where $\text{supp}(\mu) := \{x \in \mathbb{R} : \mu (x) > 0\}$).

We call μ a fuzzy real number. Denote the set of all μ with $\mathbb{R}_{\mathcal{F}}$.

E.g. $\chi_{\{x_0\}} \in \mathbb{R}_{\mathcal{F}}$, for any $x_0 \in \mathbb{R}$, where $\chi_{\{x_0\}}$ is the characteristic function at x_0.

For $0 < r \leq 1$ and $\mu \in \mathbb{R}_{\mathcal{F}}$ define

$$[\mu]^r := \{x \in \mathbb{R} : \mu(x) \geq r\}$$

and

$$[\mu]^0 := \overline{\{x \in \mathbb{R} : \mu(x) \geq 0\}}.$$

Then it is well known that for each $r \in [0,1]$, $[\mu]^r$ is a closed and bounded interval on $\mathbb{R}$ ([34]).

For $u, v \in \mathbb{R}_{\mathcal{F}}$ and $\lambda \in \mathbb{R}$, we define uniquely the sum $u \oplus v$ and the product $\lambda \odot u$ by

$$[u \oplus v]^r = [u]^r + [v]^r, \quad [\lambda \odot u]^r = \lambda [u]^r, \quad \forall\, r \in [0,1],$$

where $[u]^r + [v]^r$ means the usual addition of two intervals (as subsets of $\mathbb{R}$) and $\lambda [u]^r$ means the usual product between a scalar and a subset of $\mathbb{R}$ (see, e.g. [42]).

Notice $1 \odot u = u$ and it holds

$$u \oplus v = v \oplus u, \; \lambda \odot u = u \odot \lambda.$$

If $0 \leq r_1 \leq r_2 \leq 1$ then

$$[u]^{r_2} \subseteq [u]^{r_1}.$$

Actually $[u]^r = \left[u_-^{(r)}, u_+^{(r)}\right]$, where $u_-^{(r)} \leq u_+^{(r)}$, $u_-^{(r)}$, $u_+^{(r)} \in \mathbb{R}$, $\forall\, r \in [0,1]$.

For $\lambda > 0$ one has $\lambda u_\pm^{(r)} = (\lambda \odot u)_\pm^{(r)}$, respectively.

Define $D : \mathbb{R}_{\mathcal{F}} \times \mathbb{R}_{\mathcal{F}} \to \mathbb{R}_{\mathcal{F}}$ by

$$D(u,v) := \sup_{r \in [0,1]} \max\left\{\left|u_-^{(r)} - v_-^{(r)}\right|, \left|u_+^{(r)} - v_+^{(r)}\right|\right\},$$

where

$$[v]^r = \left[v_-^{(r)}, v_+^{(r)}\right]; \quad u, v \in \mathbb{R}_{\mathcal{F}}.$$

We have that D is a metric on $\mathbb{R}_{\mathcal{F}}$.

Then $(\mathbb{R}_{\mathcal{F}}, D)$ is a complete metric space, see [42], [43].

Here $\sum^{*}$ stands for fuzzy summation and $\widetilde{o} := \chi_{\{0\}} \in \mathbb{R}_{\mathcal{F}}$ is the neural element with respect to $\oplus$, i.e.,

$$u \oplus \widetilde{0} = \widetilde{0} \oplus u = u, \quad \forall\, u \in \mathbb{R}_{\mathcal{F}}.$$

Denote

$$D^*(f,g) = \sup_{x \in X \subseteq \mathbb{R}} D(f,g),$$

where $f, g : X \to \mathbb{R}_{\mathcal{F}}$.

We mention

Definition 11.9. Let $f : X \subseteq \mathbb{R} \to \mathbb{R}_{\mathcal{F}}$, X interval, we define the (first) fuzzy modulus of continuity of f by

$$\omega_1^{(\mathcal{F})}(f,\delta)_X = \sup_{x,y \in X,\ |x-y| \le \delta} D(f(x), f(y)), \quad \delta > 0.$$

When $g : X \subseteq \mathbb{R} \to \mathbb{R}$, we define

$$\omega_1(g,\delta)_X = \sup_{x,y \in X,\ |x-y| \le \delta} |g(x) - g(y)|.$$

We define by $C_{\mathcal{F}}^{U}(\mathbb{R})$ the space of fuzzy uniformly continuous functions from $\mathbb{R} \to \mathbb{R}_{\mathcal{F}}$, also $C_{\mathcal{F}}(\mathbb{R})$ is the space of fuzzy continuous functions on $\mathbb{R}$, and $C_b(\mathbb{R}, \mathbb{R}_{\mathcal{F}})$ is the fuzzy continuous and bounded functions.

We mention

Proposition 11.1. *([7]) Let $f \in C_{\mathcal{F}}^{U}(X)$. Then $\omega_1^{(\mathcal{F})}(f,\delta)_X < \infty$, for any $\delta > 0$.*

By [11], p. 129 we have that $C_{\mathcal{F}}^{U}([a,b]) = C_{\mathcal{F}}([a,b])$, fuzzy continuous functions on $[a,b] \subset \mathbb{R}$.

Proposition 11.2. *([7]) It holds*

$$\lim_{\delta \to 0} \omega_1^{(\mathcal{F})}(f,\delta)_X = \omega_1^{(\mathcal{F})}(f,0)_X = 0,$$

iff $f \in C_{\mathcal{F}}^{U}(X)$.

Proposition 11.3. *([7]) Here $[f]^r = \left[f_-^{(r)}, f_+^{(r)}\right]$, $r \in [0,1]$. Let $f \in C_{\mathcal{F}}(\mathbb{R})$. Then $f_{\pm}^{(r)}$ are equicontinuous with respect to $r \in [0,1]$ over $\mathbb{R}$, respectively in $\pm$.*

Note. It is clear by Propositions 11.2, 11.3, that if $f \in C_{\mathcal{F}}^{U}(\mathbb{R})$, then $f_{\pm}^{(r)} \in C_U(\mathbb{R})$ (uniformly continuous on $\mathbb{R}$). Also if $f \in C_b(\mathbb{R}, \mathbb{R}_{\mathcal{F}})$ implies $f_{\pm}^{(r)} \in C_b(\mathbb{R})$ (continuous and bounded functions on $\mathbb{R}$).

Proposition 11.4. *Let $f : \mathbb{R} \to \mathbb{R}_{\mathcal{F}}$. Assume that $\omega_1^{\mathcal{F}}(f,\delta)_X$, $\omega_1\left(f_-^{(r)},\delta\right)_X$, $\omega_1\left(f_+^{(r)},\delta\right)_X$ are finite for any $\delta > 0$, $r \in [0,1]$, where X any interval of $\mathbb{R}$. Then*

$$\omega_1^{(\mathcal{F})}(f,\delta)_X = \sup_{r \in [0,1]} \max\left\{\omega_1\left(f_-^{(r)},\delta\right)_X, \omega_1\left(f_+^{(r)},\delta\right)_X\right\}.$$

Proof. Similar to Proposition 14.15, p. 246 of [11]. □

We need

Remark 11.15. ([4]). Here $r \in [0,1]$, $x_i^{(r)}, y_i^{(r)} \in \mathbb{R}$, $i = 1, ..., m \in \mathbb{N}$. Suppose that

$$\sup_{r \in [0,1]} \max\left(x_i^{(r)}, y_i^{(r)}\right) \in \mathbb{R}, \text{ for } i = 1, ..., m.$$

Then one sees easily that

$$\sup_{r \in [0,1]} \max\left(\sum_{i=1}^{m} x_i^{(r)}, \sum_{i=1}^{m} y_i^{(r)}\right) \le \sum_{i=1}^{m} \sup_{r \in [0,1]} \max\left(x_i^{(r)}, y_i^{(r)}\right). \quad (11.233)$$

We need

Definition 11.10. Let $x, y \in \mathbb{R}_{\mathcal{F}}$. If there exists $z \in \mathbb{R}_{\mathcal{F}} : x = y \oplus z$, then we call z the H-difference on x and y, denoted $x - y$.

Definition 11.11. ([41]) Let $T := [x_0, x_0 + \beta] \subset \mathbb{R}$, with $\beta > 0$. A function $f : T \to \mathbb{R}_{\mathcal{F}}$ is H-difference at $x \in T$ if there exists an $f'(x) \in \mathbb{R}_{\mathcal{F}}$ such that the limits (with respect to D)

$$\lim_{h \to 0+} \frac{f(x+h) - f(x)}{h}, \quad \lim_{h \to 0+} \frac{f(x) - f(x-h)}{h} \tag{11.234}$$

exist and are equal to $f'(x)$.

We call f' the H-derivative or fuzzy derivative of f at x.

Above is assumed that the H-differences $f(x+h) - f(x)$, $f(x) - f(x-h)$ exists in $\mathbb{R}_{\mathcal{F}}$ in a neighborhood of x.

Higher order H-fuzzy derivatives are defined the obvious way, like in the real case.

We denote by $C_{\mathcal{F}}^N(\mathbb{R})$, $N \geq 1$, the space of all N-times continuously H-fuzzy differentiable functions from $\mathbb{R}$ into $\mathbb{R}_{\mathcal{F}}$, similarly is defined $C_{\mathcal{F}}^N([a,b])$, $[a,b] \subset \mathbb{R}$.

We mention

Theorem 11.21. *([35]) Let $f : \mathbb{R} \to \mathbb{R}_{\mathcal{F}}$ be H-fuzzy differentiable. Let $t \in \mathbb{R}$, $0 \leq r \leq 1$. Clearly*

$$[f(t)]^r = \left[f(t)_-^{(r)}, f(t)_+^{(r)} \right] \subseteq \mathbb{R}.$$

Then $(f(t))_{\pm}^{(r)}$ are differentiable and

$$[f'(t)]^r = \left[\left(f(t)_-^{(r)} \right)', \left(f(t)_+^{(r)} \right)' \right].$$

I.e.

$$(f')_{\pm}^{(r)} = \left(f_{\pm}^{(r)} \right)', \quad \forall\, r \in [0,1].$$

Remark 11.16. ([6]) Let $f \in C_{\mathcal{F}}^N(\mathbb{R})$, $N \geq 1$. Then by Theorem 11.21 we obtain

$$\left[f^{(i)}(t) \right]^r = \left[\left(f(t)_-^{(r)} \right)^{(i)}, \left(f(t)_+^{(r)} \right)^{(i)} \right],$$

for $i = 0, 1, 2, ..., N$, and in particular we have that

$$\left(f^{(i)} \right)_{\pm}^{(r)} = \left(f_{\pm}^{(r)} \right)^{(i)},$$

for any $r \in [0,1]$, all $i = 0, 1, 2, ..., N$.

Note. ([6]) Let $f \in C_{\mathcal{F}}^N(\mathbb{R})$, $N \geq 1$. Then by Theorem 11.21 we have $f_{\pm}^{(r)} \in C^N(\mathbb{R})$, for any $r \in [0,1]$.

The last three items are valid also on $[a, b]$.

By [11], p. 131, if $f \in C_{\mathcal{F}}([a, b])$, then f is a fuzzy bounded function.

For the definition of general fuzzy integral we follow [36] next.

Definition 11.12. Let (Ω, Σ, μ) be a complete σ-finite measure space. We call $F : \Omega \to R_{\mathcal{F}}$ measurable iff $\forall$ closed $B \subseteq \mathbb{R}$ the function $F^{-1}(B) : \Omega \to [0, 1]$ defined by

$$F^{-1}(B)(w) := \sup_{x \in B} F(w)(x), \text{ all } w \in \Omega$$

is measurable, see [36].

Theorem 11.22. *([36]) For $F : \Omega \to \mathbb{R}_{\mathcal{F}}$,*

$$F(w) = \left\{ \left(F_{-}^{(r)}(w), F_{+}^{(r)}(w) \right) | 0 \le r \le 1 \right\},$$

the following are equivalent

(1) F is measurable,

(2) $\forall\, r \in [0, 1]$, $F_{-}^{(r)}$, $F_{+}^{(r)}$ are measurable.

Following [36], given that for each $r \in [0, 1]$, $F_{-}^{(r)}$, $F_{+}^{(r)}$ are integrable we have that the parameterized representation

$$\left\{ \left(\int_A F_{-}^{(r)} d\mu, \int_A F_{+}^{(r)} d\mu \right) | 0 \le r \le 1 \right\} \tag{11.235}$$

is a fuzzy real number for each $A \in \Sigma$.

The last fact leads to

Definition 11.13. ([36]) A measurable function $F : \Omega \to \mathbb{R}_{\mathcal{F}}$,

$$F(w) = \left\{ \left(F_{-}^{(r)}(w), F_{+}^{(r)}(w) \right) | 0 \le r \le 1 \right\}$$

is integrable if for each $r \in [0, 1]$, $F_{\pm}^{(r)}$ are integrable, or equivalently, if $F_{\pm}^{(0)}$ are integrable.

In this case, the fuzzy integral of F over $A \in \Sigma$ is defined by

$$\int_A F d\mu := \left\{ \left(\int_A F_{-}^{(r)} d\mu, \int_A F_{+}^{(r)} d\mu \right) | 0 \le r \le 1 \right\}.$$

By [36], F is integrable iff $w \to \|F(w)\|_{\mathcal{F}}$ is real-valued integrable.

Here denote

$$\|u\|_{\mathcal{F}} := D\left(u, \widetilde{0}\right), \quad \forall\, u \in \mathbb{R}_{\mathcal{F}}.$$

We need also

Theorem 11.23. *([36]) Let $F, G : \Omega \to \mathbb{R}_{\mathcal{F}}$ be integrable. Then*

(1) Let $a, b \in \mathbb{R}$, then $aF + bG$ is integrable and for each $A \in \Sigma$,

$$\int_A (aF + bG)\, d\mu = a \int_A F d\mu + b \int_A G d\mu;$$

(2) $D(F,G)$ is a real-valued integrable function and for each $A \in \Sigma$,

$$D\left(\int_A F d\mu, \int_A G d\mu\right) \leq \int_A D(F,G)\, d\mu.$$

In particular,

$$\left\| \int_A F d\mu \right\|_{\mathcal{F}} \leq \int_A \|F\|_{\mathcal{F}}\, d\mu.$$

Above μ could be the Lebesgue measure, with all the basic properties valid here too.

Basically here we have

$$\left[\int_A F d\mu\right]^r = \left[\int_A F_-^{(r)} d\mu, \int_A F_+^{(r)} d\mu\right], \tag{11.236}$$

i.e.

$$\left(\int_A F d\mu\right)_{\pm}^{(r)} = \int_A F_{\pm}^{(r)} d\mu, \quad \forall\, r \in [0,1]. \tag{11.237}$$

We need

Definition 11.14. ([13]) Let $f \in C_{\mathcal{F}}([a,b])$ (fuzzy continuous on $[a,b] \subset \mathbb{R}$), $\nu > 0$.

We define the Fuzzy Fractional left Riemann-Liouville operator as

$$J_a^{\nu} f(x) := \frac{1}{\Gamma(\nu)} \odot \int_a^x (x-t)^{\nu-1} \odot f(t)\, dt, \quad x \in [a,b], \tag{11.238}$$

$$J_a^0 f := f.$$

Also, we define the Fuzzy Fractional right Riemann-Liouville operator as

$$I_{b-}^{\nu} f(x) := \frac{1}{\Gamma(\nu)} \odot \int_x^b (t-x)^{\nu-1} \odot f(t)\, dt, \quad x \in [a,b], \tag{11.239}$$

$$I_{b-}^0 f := f.$$

We need

Definition 11.15. ([13]) We define the Fuzzy Fractional left Caputo derivative, $x \in [a,b]$.

Let $f \in C_{\mathcal{F}}^n([a,b])$, $n = \lceil \nu \rceil$, $\nu > 0$ ($\lceil \cdot \rceil$ denotes the ceiling). We define

$$D_{*a}^{\nu\mathcal{F}} f(x) := \frac{1}{\Gamma(n-\nu)} \odot \int_a^x (x-t)^{n-\nu-1} \odot f^{(n)}(t)\, dt \tag{11.240}$$

$$= \left\{ \left(\frac{1}{\Gamma(n-\nu)} \int_a^x (x-t)^{n-\nu-1} \left(f^{(n)}\right)_-^{(r)}(t)\, dt, \right.\right.$$

$$\left.\left. \frac{1}{\Gamma(n-\nu)} \int_a^x (x-t)^{n-\nu-1} \left(f^{(n)}\right)_+^{(r)}(t)\, dt \right) \middle| 0 \leq r \leq 1 \right\}$$

$$= \left\{ \left(\frac{1}{\Gamma(n-\nu)} \int_a^x (x-t)^{n-\nu-1} \left(f_-^{(r)}\right)^{(n)} (t)\, dt, \right.\right.$$

$$\left.\left. \frac{1}{\Gamma(n-\nu)} \int_a^x (x-t)^{n-\nu-1} \left(f_+^{(r)}\right)^{(n)} (t)\, dt \right) | 0 \le r \le 1 \right\}. \tag{11.241}$$

So, we get

$$\left[D_{*a}^{\nu\mathcal{F}} f(x)\right]^r = \left[\left(\frac{1}{\Gamma(n-\nu)} \int_a^x (x-t)^{n-\nu-1} \left(f_-^{(r)}\right)^{(n)} (t)\, dt, \right.\right.$$

$$\left.\left. \frac{1}{\Gamma(n-\nu)} \int_a^x (x-t)^{n-\nu-1} \left(f_+^{(r)}\right)^{(n)} (t)\, dt \right) \right], \quad 0 \le r \le 1. \tag{11.242}$$

That is

$$\left(D_{*a}^{\nu\mathcal{F}} f(x)\right)_\pm^{(r)} = \frac{1}{\Gamma(n-\nu)} \int_a^x (x-t)^{n-\nu-1} \left(f_\pm^{(r)}\right)^{(n)} (t)\, dt = \left(D_{*a}^{\nu} \left(f_\pm^{(r)}\right)\right)(x),$$

see [10], [29].

I.e. we get that

$$\left(D_{*a}^{\nu\mathcal{F}} f(x)\right)_\pm^{(r)} = \left(D_{*a}^{\nu} \left(f_\pm^{(r)}\right)\right)(x), \tag{11.243}$$

$\forall\, x \in [a,b]$, in short

$$\left(D_{*a}^{\nu\mathcal{F}} f\right)_\pm^{(r)} = D_{*a}^{\nu} \left(f_\pm^{(r)}\right), \quad \forall\, r \in [0,1]. \tag{11.244}$$

We need

Lemma 11.3. *([13]) $D_{*a}^{\nu\mathcal{F}} f(x)$ is fuzzy continuous in $x \in [a,b]$.*

We need

Definition 11.16. ([13]) We define the Fuzzy Fractional right Caputo derivative, $x \in [a,b]$.

Let $f \in C_{\mathcal{F}}^n([a,b])$, $n = \lceil \nu \rceil$, $\nu > 0$. We define

$$D_{b-}^{\nu\mathcal{F}} f(x) := \frac{(-1)^n}{\Gamma(n-\nu)} \odot \int_x^b (t-x)^{n-\nu-1} \odot f^{(n)}(t)\, dt$$

$$= \left\{ \left(\frac{(-1)^n}{\Gamma(n-\nu)} \int_x^b (t-x)^{n-\nu-1} \left(f^{(n)}\right)_-^{(r)} (t)\, dt, \right.\right.$$

$$\left.\left. \frac{(-1)^n}{\Gamma(n-\nu)} \int_x^b (t-x)^{n-\nu-1} \left(f^{(n)}\right)_+^{(r)} (t)\, dt \right) | 0 \le r \le 1 \right\} \tag{11.245}$$

$$= \left\{ \left(\frac{(-1)^n}{\Gamma(n-\nu)} \int_x^b (t-x)^{n-\nu-1} \left(f_-^{(r)}\right)^{(n)} (t)\, dt, \right.\right.$$

$$\left.\frac{(-1)^n}{\Gamma(n-\nu)}\int_x^b (t-x)^{n-\nu-1}\left(f_+^{(r)}\right)^{(n)}(t)\,dt\right)|0\le r\le 1\Bigg\}.$$

We get

$$\left[D_{b-}^{\nu\mathcal{F}}f(x)\right]^r=\left[\left(\frac{(-1)^n}{\Gamma(n-\nu)}\int_x^b (t-x)^{n-\nu-1}\left(f_-^{(r)}\right)^{(n)}(t)\,dt,\right.\right.$$

$$\left.\left.\frac{(-1)^n}{\Gamma(n-\nu)}\int_x^b (t-x)^{n-\nu-1}\left(f_+^{(r)}\right)^{(n)}(t)\,dt\right)\right],\quad 0\le r\le 1.$$

That is

$$\left(D_{b-}^{\nu\mathcal{F}}f(x)\right)_\pm^{(r)}=\frac{(-1)^n}{\Gamma(n-\nu)}\int_x^b (t-x)^{n-\nu-1}\left(f_\pm^{(r)}\right)^{(n)}(t)\,dt=\left(D_{b-}^{\nu}\left(f_\pm^{(r)}\right)\right)(x),$$

see [9].

I.e. we get that

$$\left(D_{b-}^{\nu\mathcal{F}}f(x)\right)_\pm^{(r)}=\left(D_{b-}^{\nu}\left(f_\pm^{(r)}\right)\right)(x),\tag{11.246}$$

$\forall\, x\in[a,b]$, in short

$$\left(D_{b-}^{\nu\mathcal{F}}f\right)_\pm^{(r)}=D_{b-}^{\nu}\left(f_\pm^{(r)}\right),\quad \forall\, r\in[0,1].\tag{11.247}$$

Clearly,

$$D_{b-}^{\nu}\left(f_-^{(r)}\right)\le D_{b-}^{\nu}\left(f_+^{(r)}\right),\quad \forall\, r\in[0,1].$$

We need

Lemma 11.4. *([13]) $D_{b-}^{\nu\mathcal{F}}f(x)$ is fuzzy continuous in $x\in[a,b]$.*

11.2.6 *Fuzzy and Fuzzy-Fractional Univariate Neural Network Approximation and Interpolation*

We give

Definition 11.17. Let $f\in C_{\mathcal{F}}([a,b])$. We set

$$\left(H_n^{\mathcal{F}}(f)\right)(x):=\frac{\sum_{k=0}^{n^*} f(x_k)\odot B\left(\frac{Tn(x-x_k)}{b-a}\right)}{\sum_{k=0}^{n} B\left(\frac{Tn(x-x_k)}{b-a}\right)},\tag{11.248}$$

and we call it fuzzy interpolation univariate Neural Network operator.

Comment

We observe that

$$\left[\left(H_n^{\mathcal{F}}(f)\right)(x)\right]^r=\sum_{k=0}^{n}\left[f\left(x_k\right)\right]^r\frac{B\left(\frac{Tn(x-x_k)}{b-a}\right)}{V(x)}$$

$$=\sum_{k=0}^{n}\left[f_-^{(r)}\left(x_k\right),f_+^{(r)}\left(x_k\right)\right]\frac{B\left(\frac{Tn(x-x_k)}{b-a}\right)}{V(x)}$$

$$=\left[\sum_{k=0}^{n}f_-^{(r)}\left(x_k\right)\frac{B\left(\frac{Tn(x-x_k)}{b-a}\right)}{V(x)},\sum_{k=0}^{n}f_+^{(r)}\left(x_k\right)\frac{B\left(\frac{Tn(x-x_k)}{b-a}\right)}{V(x)}\right]$$

$$=\left[\left(H_n\left(f_-^{(r)}\right)\right)(x),\left(H_n\left(f_+^{(r)}\right)\right)(x)\right]. \tag{11.249}$$

We have proved that

$$\left(H_n^{\mathcal{F}}(f)\right)_{\pm}^{(r)}=H_n\left(f_{\pm}^{(r)}\right), \tag{11.250}$$

$\forall\, r\in[0,1]$, respectively.

Comment

We notice also that

$$\left(\left(H_n^{\mathcal{F}}(f)\right)(x_i)\right)_{\pm}^{(r)}=\left(H_n\left(f_{\pm}^{(r)}\right)\right)(x_i)=f_{\pm}^{(r)}(x_i),\ \ i=0,1,...,n,\ \forall\, r\in[0,1]. \tag{11.251}$$

Remark 11.17 (Conclusion). *(by [34], [36])*

$$\left(H_n^{\mathcal{F}}(f)\right)(x_i)=f(x_i),\ \ i=0,1,...,n,$$

the interpolation property is true at fuzzy setting.

We make

Remark 11.18. Let $f\in C_{\mathcal{F}}([a,b])$. We notice that

$$D\left(\left(H_n^{\mathcal{F}}(f)\right)(x),f(x)\right)$$

$$=\sup_{r\in[0,1]}\max\left\{\left|\left(H_n(f)\right)_-^{(r)}(x)-f_-^{(r)}(x)\right|,\left|\left(H_n(f)\right)_+^{(r)}(x)-f_+^{(r)}(x)\right|\right\}$$

$$=\sup_{r\in[0,1]}\max\left\{\left|\left(H_n\left(f_-^{(r)}\right)\right)(x)-f_-^{(r)}(x)\right|,\left|\left(H_n\left(f_+^{(r)}\right)\right)(x)-f_+^{(r)}(x)\right|\right\} \tag{11.252}$$

(hence $f_{\pm}^{(r)}\in C([a,b])$)

$$\leq\frac{2B^*}{B\left(\frac{T}{2}\right)}\sup_{r\in[0,1]}\max\left\{\omega_1\left(f_-^{(r)},\frac{b-a}{n}\right),\omega_1\left(f_+^{(r)},\frac{b-a}{n}\right)\right\}$$

(by Theorem 11.5 and Proposition 11.4)

$$=\frac{2B^*}{B\left(\frac{T}{2}\right)}\omega_1^{(\mathcal{F})}\left(f,\frac{b-a}{n}\right). \tag{11.253}$$

We have proved that

Theorem 11.24. *Let $f \in C_{\mathcal{F}}([a,b])$, $x \in [a,b]$. Then*

1)

$$D\left(\left(H_n^{\mathcal{F}}(f)\right)(x), f(x)\right) \leq \frac{2B^*}{B\left(\frac{T}{2}\right)} \omega_1^{(\mathcal{F})}\left(f, \frac{b-a}{n}\right), \tag{11.254}$$

so that $\left(H_n^{\mathcal{F}}(f)\right)(x) \xrightarrow{D} f(x)$, as $n \to \infty$, pointwise,

and

2)

$$D^*\left(H_n^{\mathcal{F}}(f), f\right) \leq \frac{2B^*}{B\left(\frac{T}{2}\right)} \omega_1^{(\mathcal{F})}\left(f, \frac{b-a}{n}\right), \tag{11.255}$$

so that $H_n^{\mathcal{F}}(f) \xrightarrow{D^*} f$, as $n \to \infty$, uniformly.

Taking into account fuzzy smoothness of f we give

Theorem 11.25. *Let $f \in C_{\mathcal{F}}^N([a,b])$, $N \in \mathbb{N}$, $x \in [a,b]$. Then*

1)

$$D\left(\left(H_n^{\mathcal{F}}(f)\right)(x), f(x)\right)$$

$$\leq \frac{2B^*}{B\left(\frac{T}{2}\right)} \left\{ \sum_{j=1}^{N} \frac{(b-a)^j}{j! n^j} D\left(f^{(j)}(x), \widetilde{o}\right) + \frac{(b-a)^N}{N! n^N} \omega_1^{(\mathcal{F})}\left(f^{(N)}, \frac{b-a}{n}\right) \right\}, \tag{11.256}$$

2) assume more that $D\left(f^{(j)}(x), \widetilde{o}\right) = 0$, $j = 1, ..., N$, where $x \in [a,b]$ is fixed, we get

$$D\left(\left(H_n^{\mathcal{F}}(f)\right)(x), f(x)\right) \leq \frac{2B^*}{B\left(\frac{T}{2}\right)} \frac{(b-a)^N}{N! n^N} \omega_1^{(\mathcal{F})}\left(f^{(N)}, \frac{b-a}{n}\right), \tag{11.257}$$

a fuzzy pointwise convergence at high speed $\frac{1}{n^{N+1}}$,

3)

$$D^*\left(H_n^{\mathcal{F}}(f), f\right)$$

$$\leq \frac{2B^*}{B\left(\frac{T}{2}\right)} \left\{ \sum_{j=1}^{N} \frac{(b-a)^j}{j! n^j} D^*\left(f^{(j)}, \widetilde{o}\right) + \frac{(b-a)^N}{N! n^N} \omega_1^{(\mathcal{F})}\left(f^{(N)}, \frac{b-a}{n}\right) \right\}. \tag{11.258}$$

Proof. Here clearly $f_{\pm}^{(r)} \in C^N([a,b])$, $\forall\, r \in [0,1]$. Then

$$D\left(\left(H_n^{\mathcal{F}}(f)\right)(x), f(x)\right)$$

$$= \sup_{r \in [0,1]} \max\left\{ \left| \left(H_n^{\mathcal{F}}(f)\right)_-^{(r)}(x) - f_-^{(r)}(x) \right|, \left| \left(H_n^{\mathcal{F}}(f)\right)_+^{(r)}(x) - f_+^{(r)}(x) \right| \right\}$$

$$= \sup_{r \in [0,1]} \max\left\{ \left| \left(H_n\left(f_-^{(r)}\right)\right)(x) - f_-^{(r)}(x) \right|, \left| \left(H_n\left(f_+^{(r)}\right)\right)(x) - f_+^{(r)}(x) \right| \right\} \tag{11.259}$$

$$\overset{(\text{by } (11.34))}{\leq} \frac{2B^*}{B\left(\frac{T}{2}\right)} \sup_{r\in[0,1]} \max \left\{ \sum_{j=1}^{N} \frac{|(f_-^{(r)})^{(j)}(x)|}{j!} \frac{(b-a)^j}{n^j} \right.$$

$$+\omega_1 \left((f_-^{(r)})^{(N)}, \frac{b-a}{n} \right) \frac{(b-a)^N}{N!n^N},$$

$$\left. \sum_{j=1}^{N} \frac{\left| \left(f_+^{(r)}\right)^{(j)}(x) \right|}{j!} \frac{(b-a)^j}{n^j} + \omega_1 \left(\left(f_+^{(r)}\right)^{(N)}, \frac{b-a}{n} \right) \frac{(b-a)^N}{N!n^N} \right\}$$

$$= \frac{2B^*}{B\left(\frac{T}{2}\right)} \sup_{r\in[0,1]} \max \left\{ \sum_{j=1}^{N} \frac{|(f_-^{(j)})^{(r)}(x)|}{j!} \frac{(b-a)^j}{n^j} + \omega_1 \left((f_-^{(N)})^{(r)}, \frac{b-a}{n} \right) \frac{(b-a)^N}{N!n^N}, \right. \tag{11.260}$$

$$\left. \sum_{j=1}^{N} \frac{\left| \left(f_+^{(j)}\right)^{(r)}(x) \right|}{j!} \frac{(b-a)^j}{n^j} + \omega_1 \left(\left(f_+^{(N)}\right)^{(r)}, \frac{b-a}{n} \right) \frac{(b-a)^N}{N!n^N} \right\}$$

$$\leq \frac{2B^*}{B\left(\frac{T}{2}\right)} \left\{ \sum_{j=1}^{N} \frac{(b-a)^j}{j!n^j} \sup_{r\in[0,1]} \max \left\{ \left| \left(f_-^{(j)}\right)^{(r)}(x) \right|, \left| \left(f_+^{(j)}\right)^{(r)}(x) \right| \right\} \right.$$

$$\left. + \frac{(b-a)^N}{N!n^N} \sup_{r\in[0,1]} \max \left\{ \omega_1 \left(\left(f_-^{(N)}\right)^{(r)}, \frac{b-a}{n} \right), \omega_1 \left(\left(f_+^{(N)}\right)^{(r)}, \frac{b-a}{n} \right) \right\} \right\} \tag{11.261}$$

$$= \frac{2B^*}{B\left(\frac{T}{2}\right)} \left\{ \sum_{j=1}^{N} \frac{(b-a)^j}{j!n^j} D\left(f^{(j)}(x), \widetilde{o}\right) + \frac{(b-a)^N}{N!n^N} \omega_1^{(\mathcal{F})} \left(f^{(N)}, \frac{b-a}{n} \right) \right\},$$

proving theorem. □

The related fuzzy-fractional results follow.

Theorem 11.26. *Let* $\beta > 0$, $N = \lceil \beta \rceil$, $\beta \notin \mathbb{N}$, $f \in C_{\mathcal{F}}^{N}([a,b])$, $x \in [a,b]$. *Then*

$$D\left(\left(H_n^{\mathcal{F}}(f)\right)(x), f(x)\right)$$

$$\leq \frac{B^*}{B\left(\frac{T}{2}\right)} \left[2 \sum_{j=1}^{N-1} \frac{D\left(f^{(j)}(x), \widetilde{o}\right)}{j!} \frac{(b-a)^j}{n^j} \right. \tag{11.262}$$

$$\left. + \frac{(b-a)^\beta}{\Gamma(\beta+1) n^\beta} \left[\omega_1^{(\mathcal{F})} \left(\left(D_{x-}^{\beta\mathcal{F}} f\right), \frac{b-a}{n} \right) + \omega_1^{(\mathcal{F})} \left(\left(D_{*x}^{\beta\mathcal{F}} f\right), \frac{b-a}{n} \right) \right] \right].$$

Proof. We get that $f_{\pm}^{(r)} \in C^N([a,b])$, $\forall\, r \in [0,1]$, and $D_{x-}^{\beta\mathcal{F}} f$, $D_{*x}^{\beta\mathcal{F}} f$ are fuzzy continuous on $[a,b]$, $\forall\, x \in [a,b]$, so that $\left(D_{x-}^{\beta\mathcal{F}} f\right)_{\pm}^{(r)}$, $\left(D_{*x}^{\beta\mathcal{F}} f\right)_{\pm}^{(r)} \in C([a,b])$, $\forall$ $x \in [a,b]$, $\forall\, r \in [0,1]$. By (11.74) we get

$$\left|H_n\left(f_{\pm}^{(r)}, x\right) - f_{\pm}^{(r)}(x)\right|$$

$$\leq \frac{B^*}{B\left(\frac{T}{2}\right)}\left[2\sum_{j=1}^{N-1} \frac{\left|\left(f_{\pm}^{(r)}\right)^{(j)}(x)\right|}{j!}\frac{(b-a)^j}{n^j}\right. \tag{11.263}$$

$$\left.+\frac{(b-a)^{\beta}}{\Gamma(\beta+1)\,n^{\beta}}\left[\omega_1\left(D_{x-}^{\beta}\left(f_{\pm}^{(r)}\right), \frac{b-a}{n}\right) + \omega_1\left(D_{*x}^{\beta}\left(f_{\pm}^{(r)}\right), \frac{b-a}{n}\right)\right]\right]$$

$$= \frac{B^*}{B\left(\frac{T}{2}\right)}\left[2\sum_{j=1}^{N-1} \frac{\left|\left(f^{(j)}(x)\right)_{\pm}^{(r)}\right|}{j!}\frac{(b-a)^j}{n^j}\right.$$

$$\left.+\frac{(b-a)^{\beta}}{\Gamma(\beta+1)\,n^{\beta}}\left[\omega_1\left(\left(D_{x-}^{\beta\mathcal{F}} f\right)_{\pm}^{(r)}, \frac{b-a}{n}\right) + \omega_1\left(\left(D_{*x}^{\beta\mathcal{F}} f\right)_{\pm}^{(r)}, \frac{b-a}{n}\right)\right]\right] \tag{11.264}$$

$$\leq \frac{B^*}{B\left(\frac{T}{2}\right)}\left[2\sum_{j=1}^{N-1} \frac{D\left(f^{(j)}(x), \widetilde{o}\right)}{j!}\frac{(b-a)^j}{n^j}\right.$$

$$\left.+\frac{(b-a)^{\beta}}{\Gamma(\beta+1)\,n^{\beta}}\left[\omega_1^{(\mathcal{F})}\left(\left(D_{x-}^{\beta\mathcal{F}} f\right), \frac{b-a}{n}\right) + \omega_1^{(\mathcal{F})}\left(\left(D_{*x}^{\beta\mathcal{F}} f\right), \frac{b-a}{n}\right)\right]\right], \tag{11.265}$$

proving the claim. □

Corollary 11.2. *(to Theorem 11.26) Assume more that* $D\left(f^{(j)}(x), \widetilde{o}\right) = 0$, *for* $j = 1, ..., N-1$, *for a fixed* $x \in [a,b]$. *Then*

$$D\left(\left(H_n^{\mathcal{F}}(f)\right)(x), f(x)\right) \leq \frac{B^*}{B\left(\frac{T}{2}\right)}\frac{(b-a)^{\beta}}{\Gamma(\beta+1)\,n^{\beta}}\cdot$$

$$\left[\omega_1^{(\mathcal{F})}\left(\left(D_{x-}^{\beta\mathcal{F}} f\right), \frac{b-a}{n}\right) + \omega_1^{(\mathcal{F})}\left(\left(D_{*x}^{\beta\mathcal{F}} f\right), \frac{b-a}{n}\right)\right], \tag{11.266}$$

fuzzy pointwise convergence at high speed of $\frac{1}{n^{\beta+1}}$.

Theorem 11.27. *Let $\beta > 0$, $N = \lceil \beta \rceil$, $\beta \notin \mathbb{N}$, $f \in C_{\mathcal{F}}^{N}([a,b])$. Then*

$$D^*\left(H_n^{\mathcal{F}}(f), f\right) \leq$$

$$\frac{B^*}{B\left(\frac{T}{2}\right)}\left[2\sum_{j=1}^{N-1}\frac{D^*\left(f^{(j)},\widetilde{o}\right)}{j!}\frac{(b-a)^j}{n^j}+\frac{(b-a)^{\beta}}{\Gamma(\beta+1)\,n^{\beta}}\cdot\right.$$

$$\left.\left[\sup_{x\in[a,b]}\omega_1^{(\mathcal{F})}\left(\left(D_{x-}^{\beta\mathcal{F}}f\right),\frac{b-a}{n}\right)+\sup_{x\in[a,b]}\omega_1^{(\mathcal{F})}\left(\left(D_{*x}^{\beta\mathcal{F}}f\right),\frac{b-a}{n}\right)\right]\right]<+\infty. \tag{11.267}$$

Proof. We notice the following

$$\left(D_{x-}^{\beta\mathcal{F}}f\right)_{\pm}^{(r)}(t)=\left(D_{x-}^{\beta}\left(f_{\pm}^{(r)}\right)\right)(t)$$

$$=\frac{(-1)^N}{\Gamma(N-\beta)}\int_t^x (s-t)^{N-\beta-1}\left(f_{\pm}^{(r)}\right)^{(N)}(s)\,ds, \tag{11.268}$$

all $a \leq t \leq x$.

Hence it holds

$$\left|\left(D_{x-}^{\beta\mathcal{F}}f\right)_{\pm}^{(r)}(t)\right|\leq\frac{1}{\Gamma(N-\beta)}\int_t^x (s-t)^{N-\beta-1}\left|\left(f_{\pm}^{(r)}\right)^{(N)}(s)\right|ds$$

$$\leq\frac{\left\|\left(f^{(N)}\right)_{\pm}^{(r)}\right\|_{\infty}(b-a)^{N-\beta}}{\Gamma(N-\beta+1)}\leq\frac{D^*\left(f^{(N)},\widetilde{o}\right)}{\Gamma(N-\beta+1)}(b-a)^{N-\beta}, \tag{11.269}$$

$a \leq t \leq x$.

Thus

$$\left\|\left(D_{x-}^{\beta\mathcal{F}}f\right)_{\pm}^{(r)}\right\|_{\infty}\leq\frac{D^*\left(f^{(N)},\widetilde{o}\right)}{\Gamma(N-\beta+1)}(b-a)^{N-\beta} \tag{11.270}$$

(notice $\left(D_{x-}^{\beta\mathcal{F}}f\right)_{\pm}^{(r)}(t)=0$, for $x \leq t \leq b$), $\forall\, r \in [0,1]$.

So that

$$D^*\left(\left(D_{x-}^{\beta\mathcal{F}}f\right),\widetilde{o}\right)\leq\frac{D^*\left(f^{(N)},\widetilde{o}\right)}{\Gamma(N-\beta+1)}(b-a)^{N-\beta}. \tag{11.271}$$

Similarly we have

$$\left(D_{*x}^{\beta\mathcal{F}}f\right)_{\pm}^{(r)}(t)=\left(D_{*x}^{\beta}\left(f_{\pm}^{(r)}\right)\right)(t)$$

$$=\frac{1}{\Gamma(N-\beta)}\int_x^t (t-s)^{N-\beta-1}\left(f_{\pm}^{(r)}\right)^{(N)}(s)\,ds, \tag{11.272}$$

where $x \leq t \leq b$.

Thus

$$\left|\left(D_{*x}^{\beta\mathcal{F}}f\right)_{\pm}^{(r)}(t)\right| \leq \frac{1}{\Gamma(N-\beta)}\int_x^t (t-s)^{N-\beta-1}\left|\left(f_{\pm}^{(r)}\right)^{(N)}(s)\right| ds$$

$$= \frac{1}{\Gamma(N-\beta)}\int_x^t (t-s)^{N-\beta-1}\left|\left(f^{(N)}\right)_{\pm}^{(r)}(s)\right| ds$$

$$\leq \frac{\left\|\left(f^{(N)}\right)_{\pm}^{(r)}\right\|_{\infty}}{\Gamma(N-\beta+1)}(b-a)^{N-\beta} \tag{11.273}$$

$$\leq \frac{D^*\left(f^{(N)},\widetilde{o}\right)}{\Gamma(N-\beta+1)}(b-a)^{N-\beta}, \quad x \leq t \leq b. \tag{11.274}$$

So that

$$\left|\left(D_{*x}^{\beta\mathcal{F}}f\right)_{\pm}^{(r)}(t)\right| \leq \frac{D^*\left(f^{(N)},\widetilde{o}\right)}{\Gamma(N-\beta+1)}(b-a)^{N-\beta}, \tag{11.275}$$

$x \leq t \leq b$.

(Notice $\left(D_{*x}^{\beta\mathcal{F}}f\right)_{\pm}^{(r)}(t) = 0$, for $a \leq t \leq x$, $\forall\, r \in [0,1]$.)

Thus

$$\left\|\left(D_{*x}^{\beta\mathcal{F}}f\right)_{\pm}^{(r)}\right\|_{\infty} \leq \frac{D^*\left(f^{(N)},\widetilde{o}\right)}{\Gamma(N-\beta+1)}(b-a)^{N-\beta}, \tag{11.276}$$

$\forall\, r \in [0,1]$.

Therefore

$$D^*\left(\left(D_{*x}^{\beta\mathcal{F}}f\right),\widetilde{o}\right) \leq \frac{D^*\left(f^{(N)},\widetilde{o}\right)}{\Gamma(N-\beta+1)}(b-a)^{N-\beta}. \tag{11.277}$$

We have proved that

$$\left\{\begin{array}{l} D^*\left(\left(D_{x-}^{\beta\mathcal{F}}f\right),\widetilde{o}\right) \\ D^*\left(\left(D_{*x}^{\beta\mathcal{F}}f\right),\widetilde{o}\right) \end{array}\right. \leq \frac{D^*\left(f^{(N)},\widetilde{o}\right)}{\Gamma(N-\beta+1)}(b-a)^{N-\beta}. \tag{11.278}$$

Next we see that

$$\omega_1^{(\mathcal{F})}\left(\left(D_{x-}^{\beta\mathcal{F}}f\right),\frac{b-a}{n}\right) = \sup_{\substack{z_1,z_2\in[a,b]\\ :|z_1-z_2|\leq\frac{b-a}{n}}} D\left(\left(D_{x-}^{\beta\mathcal{F}}f\right)(z_1),\left(D_{x-}^{\beta\mathcal{F}}f\right)(z_2)\right) \tag{11.279}$$

$$\leq \sup_{\substack{z_1,z_2\in[a,b]\\ :|z_1-z_2|\leq\frac{b-a}{n}}}\left\{D\left(\left(D_{x-}^{\beta\mathcal{F}}f\right)(z_1),\widetilde{o}\right)+D\left(\left(D_{x-}^{\beta\mathcal{F}}f\right)(z_2),\widetilde{o}\right)\right\}$$

$$\leq 2D^*\left(\left(D_{x-}^{\beta\mathcal{F}}f\right),\widetilde{o}\right) \leq \frac{2D^*\left(f^{(N)},\widetilde{o}\right)}{\Gamma(N-\beta+1)}(b-a)^{N-\beta} =: \gamma < \infty. \tag{11.280}$$

Therefore it holds

$$\sup_{x\in[a,b]}\omega_1^{(\mathcal{F})}\left(\left(D_{x-}^{\beta\mathcal{F}}f\right),\frac{b-a}{n}\right) \leq \gamma < \infty. \tag{11.281}$$

Totally similar we get

$$\sup_{x\in[a,b]}\omega_1^{(\mathcal{F})}\left(\left(D_{*x}^{\beta\mathcal{F}}f\right),\frac{b-a}{n}\right) \leq \gamma < \infty. \tag{11.282}$$

Using (11.262), (11.281), (11.282) we have established (11.267). □

11.2.7 Multivariate Fuzzy Analysis Background

Let $f, g : \prod_{i=1}^{d} [a_i, b_i] \to \mathbb{R}_{\mathcal{F}}$. We define the distance

$$D^*(f, g) := \sup_{x \in \prod_{i=1}^{d} [a_i, b_i]} D(f(x), g(x)). \tag{11.283}$$

Definition 11.18. Let $f \in C\left(\prod_{i=1}^{d} [a_i, b_i]\right)$, $d \in \mathbb{N}$, we define $(h > 0)$

$$\omega_1(f, h) := \sup_{\text{all } x_i, x_i' \in [a_i, b_i], |x_i - x_i'| \leq h, \text{ for } i=1,...,d} |f(x_1, ..., x_d) - f(x_1', ..., x_d')|. \tag{11.284}$$

For convenience call $Q := \prod_{i=1}^{d} [a_i, b_i]$.

Definition 11.19. Let $f : Q \to \mathbb{R}_{\mathcal{F}}$, we define the fuzzy modulus of continuity of f by

$$\omega_1^{(\mathcal{F})}(f, \delta) = \sup_{x, y \in Q, |x_i - y_i| \leq \delta, \text{ for } i=1,...,d} D(f(x), f(y)), \quad \delta > 0, \tag{11.285}$$

where $x = (x_1, ..., x_d)$, $y = (y_1, ..., y_d)$.

For $f : Q \to \mathbb{R}_{\mathcal{F}}$, we use

$$[f]^r = \left[f_-^{(r)}, f_+^{(r)}\right], \tag{11.286}$$

where $f_\pm^{(r)} : Q \to \mathbb{R}$, $\forall\, r \in [0, 1]$.

We need

Proposition 11.5. *Let $f : Q \to \mathbb{R}_{\mathcal{F}}$. Assume that $\omega_1^{\mathcal{F}}(f, \delta)$, $\omega_1\left(f_-^{(r)}, \delta\right)$, $\omega_1\left(f_+^{(r)}, \delta\right)$ are finite for any $\delta > 0$, $r \in [0, 1]$.*

Then

$$\omega_1^{(\mathcal{F})}(f, \delta) = \sup_{r \in [0,1]} \max\left\{\omega_1\left(f_-^{(r)}, \delta\right), \omega_1\left(f_+^{(r)}, \delta\right)\right\}. \tag{11.287}$$

Proof. By [11], p. 128. □

We define $C_{\mathcal{F}}(Q)$ the space of fuzzy continuous functions on Q.

We mention

Proposition 11.6. *Let $f \in C_{\mathcal{F}}(Q)$. Then $\omega_1^{(\mathcal{F})}(f, \delta) < \infty$, for any $\delta > 0$.*

Proof. By [11], p. 129. □

Proposition 11.7. *It holds*

$$\lim_{\delta \to 0} \omega_1^{(\mathcal{F})}(f, \delta) = \omega_1^{(\mathcal{F})}(f, 0) = 0, \tag{11.288}$$

iff $f \in C_{\mathcal{F}}(Q)$.

Proof. By [11], p. 129. □

Proposition 11.8. *Let $f \in C_{\mathcal{F}}(Q)$. Then $f_{\pm}^{(r)}$ are equicontinuous with respect to $r \in [0,1]$ over Q, respectively in $\pm$. Also f is a fuzzy bounded function.*

Proof. By [11], pp. 131, 132. □

We call $C_{\mathcal{F}}^{N}(Q)$, $N \in \mathbb{N}$, the space of all N-times fuzzy continuously differentiable functions from Q into $\mathbb{R}_{\mathcal{F}}$.

Let $f \in C_{\mathcal{F}}^{N}(Q)$, denote $f_{\widetilde{\alpha}} := \frac{\partial^{\widetilde{\alpha}} f}{\partial x^{\widetilde{\alpha}}}$, where $\widetilde{\alpha} := (\alpha_1, ..., \alpha_d)$, $\alpha_i \in \mathbb{Z}^{+}$, $i = 1, ..., d$ and

$$0 < |\widetilde{\alpha}| := \sum_{i=1}^{d} \alpha_i \leq N, \;\; N > 1.$$

Then by Theorem 11.21 we get that

$$\left(f_{\pm}^{(r)}\right)_{\widetilde{\alpha}} = (f_{\widetilde{\alpha}})_{\pm}^{(r)}, \;\; \forall\, r \in [0,1], \tag{11.289}$$

and any $\widetilde{\alpha} : |\widetilde{\alpha}| \leq N$. Here $f_{\pm}^{(r)} \in C^{N}(Q)$.

Notation 11.1. We denote

$$\left(\sum_{i=1}^{2} D\left(\frac{\partial}{\partial x_i}, \widetilde{0}\right)\right)^{2} f(x)$$

$$:= D\left(\frac{\partial^2 f(x_1, x_2)}{\partial x_1^2}, \widetilde{0}\right) + D\left(\frac{\partial^2 f(x_1, x_2)}{\partial x_2^2}, \widetilde{0}\right) + 2D\left(\frac{\partial^2 f(x_1, x_2)}{\partial x_1 \partial x_2}, \widetilde{0}\right). \tag{11.290}$$

In general we denote ($j = 1, ..., N$)

$$\left(\sum_{i=1}^{d} D\left(\frac{\partial}{\partial x_i}, \widetilde{0}\right)\right)^{j} f(x)$$

$$:= \sum_{(j_1,...,j_d) \in \mathbb{Z}_{+}^{d} : \sum_{i=1}^{d} j_i = j} \frac{j!}{j_1! j_2! ... j_d!} D\left(\frac{\partial^j f(x_1, ..., x_d)}{\partial x_1^{j_1} \partial x_2^{j_2} ... \partial x_d^{j_d}}, \widetilde{0}\right). \tag{11.291}$$

Let

$$f_{\widetilde{\alpha}}(x) = \widetilde{o}, \text{ for all } \widetilde{\alpha} : |\widetilde{\alpha}| = 1, ..., N,$$

for $x \in Q$ fixed.

The last implies $D(f_{\widetilde{\alpha}}(x), \widetilde{o}) = 0$, and by (11.291) we obtain

$$\left[\left(\sum_{i=1}^{d} D\left(\frac{\partial}{\partial x_i}, \widetilde{o}\right)\right)^{j} f(x)\right] = 0, \tag{11.292}$$

for $j = 1, ..., N$.

11.2.8 *Multivariate Fuzzy Neural Network Approximation and Interpolation*

Let $f \in C_{\mathcal{F}}\left(\prod_{i=1}^{d}[a_i,b_i]\right)$, $x \in \prod_{i=1}^{d}[a_i,b_i]$, we define

$$M_n^{\mathcal{F}}(f,x) := M_n^{\mathcal{F}}(f,x_1,...,x_d)$$

$$:= \frac{\sum_{k_1=0}^{n*} ... \sum_{k_d=0}^{n*} f(x_{k_1 1},...,x_{k_d d}) \odot E\left(\frac{T_1 n(x_1 - x_{k_1 1})}{b_1 - a_1},...,\frac{T_d n(x_d - x_{k_d d})}{b_d - a_d}\right)}{\sum_{k_1=0}^{n} ... \sum_{k_d=0}^{n} E\left(\frac{T_1 n(x_1 - x_{k_1 1})}{b_1 - a_1},...,\frac{T_d n(x_d - x_{k_d d})}{b_d - a_d}\right)}, \quad (11.293)$$

the multivariate fuzzy neural network interpolation operator, $\forall\, n \in \mathbb{N}$.

Remark 11.19. We observe that

$$[M_n^{\mathcal{F}}(f,x)]^r$$

$$= \frac{\sum_{k_1=0}^{n} ... \sum_{k_d=0}^{n} [f(x_{k_1 1},...,x_{k_d d})]^r\, E\left(\frac{T_1 n(x_1 - x_{k_1 1})}{b_1 - a_1},...,\frac{T_d n(x_d - x_{k_d d})}{b_d - a_d}\right)}{W} \quad (11.294)$$

$$= \sum_{k_1=0}^{n} ... \sum_{k_d=0}^{n} \left[f_-^{(r)}(x_{k_1 1},...,x_{k_d d}), f_+^{(r)}(x_{k_1 1},...,x_{k_d d})\right] \cdot$$

$$\frac{E\left(\frac{T_1 n(x_1 - x_{k_1 1})}{b_1 - a_1},...,\frac{T_d n(x_d - x_{k_d d})}{b_d - a_d}\right)}{W}$$

$$= \left[\sum_{k_1=0}^{n} ... \sum_{k_d=0}^{n} f_-^{(r)}(x_{k_1 1},...,x_{k_d d}) \frac{E(>>)}{W},\right.$$

$$\left.\sum_{k_1=0}^{n} ... \sum_{k_d=0}^{n} f_+^{(r)}(x_{k_1 1},...,x_{k_d d}) \frac{E(>>)}{W}\right] \quad (11.295)$$

$$= \left[\left(M_n\left(f_-^{(r)}\right)\right)(x), \left(M_n\left(f_+^{(r)}\right)\right)(x)\right].$$

Hence it holds

$$\left(M_n^{\mathcal{F}}(f)\right)_{\pm}^{(r)} = M_n\left(f_{\pm}^{(r)}\right), \quad (11.296)$$

$\forall\, r \in [0,1]$, respectively.

Remark 11.20. Let $(k_1, ..., k_d) \in \{0, 1, ..., n\}^d$. Then

$$\left(M_n^{\mathcal{F}}\left(f, x_{k_1 1}, ..., x_{k_d d}\right)\right)_{\pm}^{(r)} = M_n\left(f_{\pm}^{(r)}\right)\left(x_{k_1 1}, ..., x_{k_d d}\right) \tag{11.297}$$

$$= f_{\pm}^{(r)}\left(x_{k_1 1}, ..., x_{k_d d}\right), \quad \forall\, r \in [0,1],$$

proving

$$M_n^{\mathcal{F}}\left(f, x_{k_1 1}, ..., x_{k_d d}\right) = f\left(x_{k_1 1}, ..., x_{k_d d}\right), \tag{11.298}$$

the interpolation property.

Remark 11.21. Let $f \in C_{\mathcal{F}}\left(\prod_{i=1}^{d}[a_i, b_i]\right)$. Then

$$D\left(\left(M_n^{\mathcal{F}}(f)\right)(x), f(x)\right)$$

$$= \sup_{r\in[0,1]} \max\left\{\left|\left(M_n^{\mathcal{F}}(f)\right)_{-}^{(r)}(x) - f_{-}^{(r)}(x)\right|, \left|\left(M_n^{\mathcal{F}}(f)\right)_{+}^{(r)}(x) - f_{+}^{(r)}(x)\right|\right\}$$

$$= \sup_{r\in[0,1]} \max\left\{\left|\left(M_n^{\mathcal{F}}\left(f_{-}^{(r)}\right)\right)(x) - f_{-}^{(r)}(x)\right|, \left|\left(M_n^{\mathcal{F}}\left(f_{+}^{(r)}\right)\right)(x) - f_{+}^{(r)}(x)\right|\right\} \tag{11.299}$$

$\left(\text{we have } f_{\pm}^{r} \in C\left(\prod_{i=1}^{d}[a_i, b_i]\right)\right)$

$$\overset{(11.148)}{\leq} \sup_{r\in[0,1]} \max\left\{\frac{2^d E^*}{E\left(\frac{T_1}{2}, ..., \frac{T_d}{2}\right)}\omega_1\left(f_{-}^{(r)}, \frac{\|b-a\|_\infty}{n}\right), \right.$$

$$\left.\frac{2^d E^*}{E\left(\frac{T_1}{2}, ..., \frac{T_d}{2}\right)}\omega_1\left(f_{+}^{(r)}, \frac{\|b-a\|_\infty}{n}\right)\right\} \tag{11.300}$$

$$= \frac{2^d E^*}{E\left(\frac{T_1}{2}, ..., \frac{T_d}{2}\right)} \sup_{r\in[0,1]} \max\left\{\omega_1\left(f_{-}^{(r)}, \frac{\|b-a\|_\infty}{n}\right), \omega_1\left(f_{+}^{(r)}, \frac{\|b-a\|_\infty}{n}\right)\right\} \tag{11.301}$$

$$\overset{(11.287)}{=} \frac{2^d E^*}{E\left(\frac{T_1}{2}, ..., \frac{T_d}{2}\right)}\omega_1^{(\mathcal{F})}\left(f, \frac{\|b-a\|_\infty}{n}\right).$$

We have proved

Theorem 11.28. *Let* $f \in C_{\mathcal{F}}\left(\prod_{i=1}^{d}[a_i, b_i]\right)$. *Then*

$$D\left(\left(M_n^{\mathcal{F}}(f)\right)(x), f(x)\right)$$

$$\leq \frac{2^d E^*}{E\left(\frac{T_1}{2}, ..., \frac{T_d}{2}\right)}\omega_1^{(\mathcal{F})}\left(f, \frac{\|b-a\|_\infty}{n}\right) =: \lambda, \tag{11.302}$$

and

$$D^*\left(M_n^{\mathcal{F}}(f), f\right) \leq \lambda. \tag{11.303}$$

We make

Remark 11.22. Let $f \in C_{\mathcal{F}}^{N}\left(\prod_{i=1}^{d}[a_i,b_i]\right)$, $N \in \mathbb{N}$, $x \in \prod_{i=1}^{d}[a_i,b_i]$ (so that $f_{\pm}^{(r)} \in C^{N}\left(\prod_{i=1}^{d}[a_i,b_i]\right)$).

We get

$$\left|M_n\left(f_{\pm}^{(r)},x\right)-f_{\pm}^{(r)}(x)\right|$$

$$\overset{(11.156)}{\leq} \frac{2^d E^*}{E\left(\frac{T_1}{2},\ldots,\frac{T_d}{2}\right)}\left[\sum_{j=1}^{N}\frac{1}{j!}\left(\frac{\|b-a\|_{\infty}^{j}}{n^j}\right)\left(\left(\sum_{i=1}^{d}\left|\frac{\partial}{\partial x_i}\right|\right)^{j} f_{\pm}^{(r)}(x)\right)\right. \tag{11.304}$$

$$\left.+\frac{\|b-a\|_{\infty}^{N} d^N}{N!n^N}\max_{\widetilde{\alpha}:|\widetilde{\alpha}|=N}\omega_1\left(\left(f_{\pm}^{(r)}\right)_{\widetilde{\alpha}},\frac{\|b-a\|_{\infty}}{n}\right)\right]$$

$$=\frac{2^d E^*}{E\left(\frac{T_1}{2},\ldots,\frac{T_d}{2}\right)}\left[\sum_{j=1}^{N}\frac{1}{j!}\left(\frac{\|b-a\|_{\infty}^{j}}{n^j}\right)\left(\left(\sum_{i=1}^{d}\left|\frac{\partial}{\partial x_i}\right|\right)^{j} f(x)\right)_{\pm}^{(r)}\right. \tag{11.305}$$

$$\left.+\frac{\|b-a\|_{\infty}^{N} d^N}{N!n^N}\max_{\widetilde{\alpha}:|\widetilde{\alpha}|=N}\omega_1\left(\left(f_{\widetilde{\alpha}}\right)_{\pm}^{(r)},\frac{\|b-a\|_{\infty}}{n}\right)\right]$$

$$\leq \frac{2^d E^*}{E\left(\frac{T_1}{2},\ldots,\frac{T_d}{2}\right)}\left[\sum_{j=1}^{N}\frac{1}{j!}\left(\frac{\|b-a\|_{\infty}^{j}}{n^j}\right)\left[\left(\sum_{i=1}^{d}D\left(\frac{\partial}{\partial x_i},\widetilde{o}\right)\right)^{j} f(x)\right]\right. \tag{11.306}$$

$$\left.+\frac{\|b-a\|_{\infty}^{N} d^N}{N!n^N}\max_{\widetilde{\alpha}:|\widetilde{\alpha}|=N}\omega_1^{(\mathcal{F})}\left(f_{\widetilde{\alpha}},\frac{\|b-a\|_{\infty}}{n}\right)\right].$$

We have proved

Theorem 11.29. *Let* $f \in C_{\mathcal{F}}^{N}\left(\prod_{i=1}^{d}[a_i,b_i]\right)$, $N \in \mathbb{N}$, $x \in \prod_{i=1}^{d}[a_i,b_i]$. *Then*

$$D\left(M_n^{\mathcal{F}}(f)(x),f(x)\right)$$

$$\leq \frac{2^d E^*}{E\left(\frac{T_1}{2},\ldots,\frac{T_d}{2}\right)}\left[\sum_{j=1}^{N}\frac{1}{j!}\left(\frac{\|b-a\|_{\infty}^{j}}{n^j}\right)\left[\left(\sum_{i=1}^{d}D\left(\frac{\partial}{\partial x_i},\widetilde{o}\right)\right)^{j} f(x)\right]\right. \tag{11.307}$$

$$\left.+\frac{\|b-a\|_{\infty}^{N} d^N}{N!n^N}\max_{\widetilde{\alpha}:|\widetilde{\alpha}|=N}\omega_1^{(\mathcal{F})}\left(f_{\widetilde{\alpha}},\frac{\|b-a\|_{\infty}}{n}\right)\right].$$

Corollary 11.3. *(to Theorem 11.29) Additionally assume that* $f_{\widetilde{\alpha}}(x) = \widetilde{o}$, *for all* $\widetilde{\alpha} : |\widetilde{\alpha}| = 1, ..., N$, *where* $x \in \prod_{i=1}^{d} [a_i, b_i]$ *is fixed.*

[Then $D(f_{\widetilde{\alpha}}(x), \widetilde{o}) = 0$, *and* $\left[\left(\sum_{i=1}^{d} D\left(\frac{\partial}{\partial x_i}, \widetilde{o}\right)\right)^j f(x)\right] = 0$, $j = 1, ..., N$.*]*

Hence

$$D\left(M_n^{\mathcal{F}}(f)(x), f(x)\right) \leq$$

$$\frac{2^d E^*}{E\left(\frac{T_1}{2}, ..., \frac{T_d}{2}\right)} \frac{\|b-a\|_{\infty}^N d^N}{N! n^N} \max_{\widetilde{\alpha}:|\widetilde{\alpha}|=N} \omega_1^{(\mathcal{F})}\left(f_{\widetilde{\alpha}}, \frac{\|b-a\|_{\infty}}{n}\right). \tag{11.308}$$

Corollary 11.4. *(to Theorem 11.29) We get*

$$D^*\left(M_n^{\mathcal{F}}(f), f\right) \leq$$

$$\frac{2^d E^*}{E\left(\frac{T_1}{2}, ..., \frac{T_d}{2}\right)} \left[\sum_{j=1}^{N} \frac{1}{j!}\left(\frac{\|b-a\|_{\infty}^j}{n^j}\right) \left\|\left(\sum_{i=1}^{d} D\left(\frac{\partial}{\partial x_i}, \widetilde{o}\right)\right)^j f(x)\right\|_{\infty}\right. \tag{11.309}$$

$$\left. + \frac{\|b-a\|_{\infty}^N d^N}{N! n^N} \max_{\widetilde{\alpha}:|\widetilde{\alpha}|=N} \omega_1^{(\mathcal{F})}\left(f_{\widetilde{\alpha}}, \frac{\|b-a\|_{\infty}}{n}\right)\right].$$

Corollary 11.5. *(to Theorem 11.29) Case of* $N = 1$. *We derive*

$$D\left(\left(M_n^{\mathcal{F}}(f)\right)(x), f(x)\right) \leq \frac{2^d E^* \|b-a\|_{\infty}}{n E\left(\frac{T_1}{2}, ..., \frac{T_d}{2}\right)} \cdot$$

$$\left[\sum_{i=1}^{d} D\left(\frac{\partial f}{\partial x_i}, \widetilde{o}\right) + d \max_{i \in \{1,...,d\}} \omega_1^{(\mathcal{F})}\left(\frac{\partial f}{\partial x_i}, \frac{\|b-a\|_{\infty}}{n}\right)\right]. \tag{11.310}$$

11.2.9 *Fuzzy-Random Analysis Background*

Define

$$D : \mathbb{R}_{\mathcal{F}} \times \mathbb{R}_{\mathcal{F}} \to \mathbb{R}_+ \cup \{0\}$$

by

$$D(u, v) := \sup_{r \in [0,1]} \max\left\{\left|u_-^{(r)} - v_-^{(r)}\right|, \left|u_+^{(r)} - v_+^{(r)}\right|\right\}, \tag{11.311}$$

where $[v]^r = \left[v_-^{(r)}, v_+^{(r)}\right]$; $u, v \in \mathbb{R}_{\mathcal{F}}$. We have that D is a metric on $\mathbb{R}_{\mathcal{F}}$. Then $(\mathbb{R}_{\mathcal{F}}, D)$ is a complete metric space, see [41], with the properties

$$D(u \oplus w, v \oplus w) = D(u, v), \quad \forall\, u, v, w \in \mathbb{R}_{\mathcal{F}},$$
$$D(k \odot u, k \odot v) = |k|\, D(u, v), \quad \forall\, u, v \in \mathbb{R}_{\mathcal{F}}, \forall\, k \in \mathbb{R},$$
$$D(u \oplus v, w \oplus e) \leq D(u, w) + D(v, e), \quad \forall\, u, v, w, e \in \mathbb{R}_{\mathcal{F}}.$$

Let $U^* := \prod_{i=1}^{d} [a_i, b_i]$, $d \in \mathbb{N}$, $f, g : U^* \to \mathbb{R}_{\mathcal{F}}$ be fuzzy real number valued functions. The distance between f, g is defined by

$$D^*(f, g) := \sup_{x \in U^*} D(f(x), g(x)).$$

On $\mathbb{R}_{\mathcal{F}}$ we define a partial order by "$\leq$": $u, v \in \mathbb{R}_{\mathcal{F}}$, $u \leq v$ iff $u_-^{(r)} \leq v_-^{(r)}$ and $u_+^{(r)} \leq v_+^{(r)}$, $\forall\, r \in [0, 1]$.

We need

Lemma 11.5. *([25]) For any $a, b \in \mathbb{R} : a \cdot b \geq 0$ and any $u \in \mathbb{R}_{\mathcal{F}}$ we have*

$$D(a \odot u, b \odot u) \leq |a - b| \cdot D(u, \widetilde{o}), \tag{11.312}$$

where $\widetilde{o} \in \mathbb{R}_{\mathcal{F}}$ is defined by $\widetilde{o} := \chi_{\{0\}}$.

Lemma 11.6. *([25])*

(i) If we denote $\widetilde{o} := \chi_{\{0\}}$, then $\widetilde{o} \in \mathbb{R}_{\mathcal{F}}$ is the neutral element with respect to $\oplus$, i.e., $u \oplus \widetilde{o} = \widetilde{o} \oplus u = u$, $\forall\, u \in \mathbb{R}_{\mathcal{F}}$.

(ii) With respect to $\widetilde{o}$, none of $u \in \mathbb{R}_{\mathcal{F}}$, $u \neq \widetilde{o}$ has opposite in $\mathbb{R}_{\mathcal{F}}$.

(iii) Let $a, b \in \mathbb{R} : a \cdot b \geq 0$, and any $u \in \mathbb{R}_{\mathcal{F}}$, we have $(a + b) \odot u = a \odot u \oplus b \odot u$. For general $a, b \in \mathbb{R}$, the above property is false.

(iv) For any $\lambda \in \mathbb{R}$ and any $u, v \in \mathbb{R}_{\mathcal{F}}$, we have $\lambda \odot (u \oplus v) = \lambda \odot u \oplus \lambda \odot v$.

(v) For any $\lambda, \mu \in \mathbb{R}$ and $u \in \mathbb{R}_{\mathcal{F}}$, we have $\lambda \odot (\mu \odot u) = (\lambda \cdot \mu) \odot u$.

(vi) If we denote $\|u\|_{\mathcal{F}} := D(u, \widetilde{o})$, $\forall\, u \in \mathbb{R}_{\mathcal{F}}$, then $\|\cdot\|_{\mathcal{F}}$ has the properties of a usual norm on $\mathbb{R}_{\mathcal{F}}$, i.e.,

$$\|u\|_{\mathcal{F}} = 0 \text{ iff } u = \widetilde{o}, \quad \|\lambda \odot u\|_{\mathcal{F}} = |\lambda| \cdot \|u\|_{\mathcal{F}},$$
$$\|u \oplus v\|_{\mathcal{F}} \leq \|u\|_{\mathcal{F}} + \|v\|_{\mathcal{F}}, \quad \|u\|_{\mathcal{F}} - \|v\|_{\mathcal{F}} \leq D(u, v). \tag{11.313}$$

Notice that $(\mathbb{R}_{\mathcal{F}}, \oplus, \odot)$ is not a linear space over $\mathbb{R}$; and consequently $(\mathbb{R}_{\mathcal{F}}, \|\cdot\|_{\mathcal{F}})$ is not a normed space.

As in Remark 4.4 ([25]) one can show easily that a sequence of operators of the form

$$L_n(f)(x) := \sum_{k=0}^{n*} f(x_{k_n}) \odot w_{n,k}(x), \quad n \in \mathbb{N}, \tag{11.314}$$

($\sum^{*}$ denotes the fuzzy summation) where $f : U^* \to \mathbb{R}_{\mathcal{F}}$, $x_{k_n} \in U^*$, $w_{n,k}(x)$ real valued weights, are linear over U^*, i.e.,

$$L_n(\lambda \odot f \oplus \mu \odot g)(x) = \lambda \odot L_n(f)(x) \oplus \mu \odot L_n(g)(x), \tag{11.315}$$

$\forall \lambda, \mu \in \mathbb{R}$, any $x \in U^*$; $f, g : U^* \to \mathbb{R}_{\mathcal{F}}$. (Proof based on Lemma 11.6 (iv).)

We further need

Definition 11.20. (see also [33], Definition 13.16, p. 654) Let $(X, \mathcal{B}, P)$ be a probability space. A fuzzy-random variable is a $\mathcal{B}$-measurable mapping $g : X \to \mathbb{R}_{\mathcal{F}}$

(i.e., for any open set $Z \subseteq \mathbb{R}_{\mathcal{F}}$, in the topology of $\mathbb{R}_{\mathcal{F}}$ generated by the metric D, we have

$$g^{-1}(Z) = \{s \in X; g(s) \in Z\} \in \mathcal{B}). \tag{11.316}$$

The set of all fuzzy-random variables is denoted by $\mathcal{L}_{\mathcal{F}}(X, \mathcal{B}, P)$. Let $g_n, g \in \mathcal{L}_{\mathcal{F}}(X, \mathcal{B}, P)$, $n \in \mathbb{N}$ and $0 < q < +\infty$. We say $g_n(s) \overset{\text{“}q\text{-mean”}}{\underset{n\to+\infty}{\rightarrow}} g(s)$ if

$$\lim_{n\to+\infty} \int_X D(g_n(s), g(s))^q P(ds) = 0. \tag{11.317}$$

Remark 11.23. (see [33], p. 654) If $f, g \in \mathcal{L}_{\mathcal{F}}(X, \mathcal{B}, P)$, let us denote $F : X \to \mathbb{R}_+ \cup \{0\}$ by $F(s) = D(f(s), g(s))$, $s \in X$. Here, F is $\mathcal{B}$-measurable, because $F = G \circ H$, where $G(u, v) = D(u, v)$ is continuous on $\mathbb{R}_{\mathcal{F}} \times \mathbb{R}_{\mathcal{F}}$, and $H : X \to \mathbb{R}_{\mathcal{F}} \times \mathbb{R}_{\mathcal{F}}$, $H(s) = (f(s), g(s))$, $s \in X$, is $\mathcal{B}$-measurable. This shows that the above convergence in q-mean makes sense.

Definition 11.21. (see [33], p. 654, Definition 13.17) Let $(T, \mathcal{T})$ be a topological space. A mapping $f : T \to \mathcal{L}_{\mathcal{F}}(X, \mathcal{B}, P)$ will be called fuzzy-random function (or fuzzy-stochastic process) on T. We denote $f(t)(s) = f(t, s)$, $t \in T$, $s \in X$.

Remark 11.24. (see [33], p. 655) Any usual fuzzy real function $f : T \to \mathbb{R}_{\mathcal{F}}$ can be identified with the degenerate fuzzy-random function $f(t, s) = f(t)$, $\forall\, t \in T$, $s \in X$.

Remark 11.25. (see [33], p. 655) Fuzzy-random functions that coincide with probability one for each $t \in T$ will be considered equivalent.

Remark 11.26. (see [33], p. 655) Let $f, g : T \to \mathcal{L}_{\mathcal{F}}(X, \mathcal{B}, P)$. Then $f \oplus g$ and $k \odot f$ are defined pointwise, i.e.,

$$(f \oplus g)(t, s) = f(t, s) \oplus g(t, s),$$
$$(k \odot f)(t, s) = k \odot f(t, s), \ \ t \in T, s \in X.$$

Definition 11.22. (see also Definition 13.18, pp. 655-656, [33]) For a fuzzy-random function $f : U^* \to \mathcal{L}_{\mathcal{F}}(X, \mathcal{B}, P)$, $d \in \mathbb{N}$, we define the (first) fuzzy-random modulus of continuity

$$\Omega_1^{(\mathcal{F})}(f, \delta)_{L^q}$$

$$= \sup\left\{\left(\int_X D^q(f(x, s), f(y, s)) P(ds)\right)^{\frac{1}{q}} : x, y \in U^*, \ \|x - y\|_{l_1} \le \delta\right\}, \tag{11.318}$$

$0 < \delta$, $1 \le q < \infty$.

Definition 11.23. (as in [22]) Here $1 \leq q < +\infty$. Let $f : U^* \to \mathcal{L}_{\mathcal{F}}(X, \mathcal{B}, P)$, $d \in \mathbb{N}$, be a fuzzy random function. We call f a (q-mean) uniformly continuous fuzzy random function over U^*, iff $\forall\ \varepsilon > 0\ \exists\ \delta > 0$:whenever $\|x - y\|_{l_1} \leq \delta$, $x, y \in U^*$, implies that

$$\int_X (D(f(x,s), f(y,s)))^q P(ds) \leq \varepsilon. \tag{11.319}$$

We denote it as $f \in C_{FR}^{U_q}(U^*)$.

Proposition 11.9. *(as in [22]) Let $f \in C_{FR}^{U_q}(U^*)$. Then $\Omega_1^{(\mathcal{F})}(f, \delta)_{L^q} < \infty$, any $\delta > 0$.*

Proposition 11.10. *(as in [22]) Let $f, g : U^* \to \mathcal{L}_{\mathcal{F}}(X, \mathcal{B}, P)$, $d \in \mathbb{N}$, be fuzzy random functions. It holds*

(i) $\Omega_1^{(\mathcal{F})}(f, \delta)_{L^q}$ is nonnegative and nondecreasing in $\delta > 0$.

(ii) $\lim_{\delta \downarrow 0} \Omega_1^{(\mathcal{F})}(f, \delta)_{L^q} = \Omega_1^{(\mathcal{F})}(f, 0)_{L^q} = 0$, iff $f \in C_{FR}^{U_q}(U^*)$.

(iii) $\Omega_1^{(\mathcal{F})}(f, \delta_1 + \delta_2)_{L^q} \leq \Omega_1^{(\mathcal{F})}(f, \delta_1)_{L^q} + \Omega_1^{(\mathcal{F})}(f, \delta_2)_{L^q}$, $\delta_1, \delta_2 > 0$.

(iv) $\Omega_1^{(\mathcal{F})}(f, n\delta)_{L^q} \leq n\Omega_1^{(\mathcal{F})}(f, \delta)_{L^q}$, $\delta > 0$, $n \in \mathbb{N}$.

(v) $\Omega_1^{(\mathcal{F})}(f, \lambda\delta)_{L^q} \leq \lceil \lambda \rceil \Omega_1^{(\mathcal{F})}(f, \delta)_{L^q} \leq (\lambda + 1)\Omega_1^{(\mathcal{F})}(f, \delta)_{L^q}$, $\lambda > 0$, $\delta > 0$, where $\lceil \cdot \rceil$ is the ceiling of the number.

(vi) $\Omega_1^{(\mathcal{F})}(f \oplus g, \delta)_{L^q} \leq \Omega_1^{(\mathcal{F})}(f, \delta)_{L^q} + \Omega_1^{(\mathcal{F})}(g, \delta)_{L^q}$, $\delta > 0$. Here $f \oplus g$ is a fuzzy random function.

(vii) $\Omega_1^{(\mathcal{F})}(f, \cdot)_{L^q}$ is continuous on $\mathbb{R}_+$, for $f \in C_{FR}^{U_q}(U^*)$.

According to [30], p. 94 we have the following

Definition 11.24. Let $(Y, \mathcal{T})$ be a topological space, with its σ-algebra of Borel sets $\mathcal{B} := \mathcal{B}(Y, \mathcal{T})$ generated by $\mathcal{T}$. If $(X, \mathcal{S})$ is a measurable space, a function $f : X \to Y$ is called measurable iff $f^{-1}(B) \in \mathcal{S}$ for all $B \in \mathcal{B}$.

By Theorem 4.1.6 of [30], p. 89 f as above is measurable iff

$$f^{-1}(C) \in \mathcal{S} \text{ for all } C \in \mathcal{T}.$$

We would need

Theorem 11.30. *(see [30], p. 95) Let $(X, \mathcal{S})$ be a measurable space and (Y, d) be a metric space. Let f_n be measurable functions from X into Y such that for all $x \in X$, $f_n(x) \to f(x)$ in Y. Then f is measurable. I.e., $\lim_{n \to \infty} f_n = f$ is measurable.*

We need also

Proposition 11.11. *Let f, g be fuzzy random variables from $\mathcal{S}$ into $\mathbb{R}_{\mathcal{F}}$. Then*

(i) Let $c \in \mathbb{R}$, then $c \odot f$ is a fuzzy random variable.

(ii) $f \oplus g$ is a fuzzy random variable.

11.2.10 *Multivariate Fuzzy Random Neural Network Approximation and Interpolation*

We need

Definition 11.25. Let here $(X, \mathcal{B}, P)$ be a probability space, $s \in X$, $n \in \mathbb{N}$, $f \in C_{\mathcal{F}R}^{U_q}\left(\prod_{i=1}^{d}[a_i, b_i]\right)$, $1 \leq q < \infty$, and $x \in \prod_{i=1}^{d}[a_i, b_i]$.

We define

$$M_n^{\mathcal{F}R}(f, x, s) := M_n^{\mathcal{F}R}(f, x_1, ..., x_d, s)$$

$$:= \frac{\sum_{k_1=0}^{n*} ... \sum_{k_d=0}^{n*} f(x_{k_1 1}, ..., x_{k_d d}, s) \odot E\left(\frac{T_1 n(x_1 - x_{k_1 1})}{b_1 - a_1}, ..., \frac{T_d n(x_d - x_{k_d d})}{b_d - a_d}\right)}{\sum_{k_1=0}^{n} ... \sum_{k_d=0}^{n} E\left(\frac{T_1 n(x_1 - x_{k_1 1})}{b_1 - a_1}, ..., \frac{T_d n(x_d - x_{k_d d})}{b_d - a_d}\right)}. \tag{11.320}$$

We make

Remark 11.27. Clearly here it holds

$$M_n^{\mathcal{F}R}(f, x_{k_1 1}, ..., x_{k_d d}, s) = \frac{f(x_{k_1 1}, ..., x_{k_d d}, s) \odot E^*}{E^*}$$

$$= f(x_{k_1 1}, ..., x_{k_d d}, s) \odot 1 = f(x_{k_1 1}, ..., x_{k_d d}, s), \tag{11.321}$$

proving the interpolation property of operators $M_n^{\mathcal{F}R}$.

We make

Remark 11.28. Let $f \in C_{\mathcal{F}R}^{U_q}\left(\prod_{i=1}^{d}[a_i, b_i]\right)$, $1 \leq q < \infty$, $x \in \prod_{i=1}^{d}[a_i, b_i]$, $n \in \mathbb{N}$. We observe that

$$D\left(M_n^{\mathcal{F}R}(f, x, s), f(x, s)\right)$$

$$= D\left(\sum_{k_1=0}^{n*} ... \sum_{k_d=0}^{n*} f(x_{k_1 1}, ..., x_{k_d d}, s) \odot \frac{E\left(\frac{T_1 n(x_1 - x_{k_1 1})}{b_1 - a_1}, ..., \frac{T_d n(x_d - x_{k_d d})}{b_d - a_d}\right)}{W},\right. \tag{11.322}$$

$$\left. f(x, s) \odot \frac{W}{W}\right)$$

$$= D\left(\sum_{k_1=0}^{n*} ... \sum_{k_d=0}^{n*} f(x_{k_1 1}, ..., x_{k_d d}, s) \odot \frac{E(>>)}{W}, \sum_{k_1=0}^{n*} ... \sum_{k_d=0}^{n*} f(x, s) \odot \frac{E(>>)}{W}\right) \tag{11.323}$$

$$\leq \frac{\sum_{k_1=0}^{n} \dots \sum_{k_d=0}^{n} E(>>)}{W} D\left(f\left(x_{k_1 1}, \dots, x_{k_d d}, s\right), f(x,s)\right).$$

So it holds

$$D\left(M_n^{\mathcal{F}R}(f,x,s), f(x,s)\right)$$
$$\leq \frac{\sum_{k_1=0}^{n} \dots \sum_{k_d=0}^{n} E\left(\frac{T_1 n(x_1 - x_{k_1 1})}{b_1 - a_1}, \dots, \frac{T_d n(x_d - x_{k_d d})}{b_d - a_d}\right)}{W} D\left(f\left(x_{k_1 1}, \dots, x_{k_d d}, s\right), f(x,s)\right). \tag{11.324}$$

Therefore we derive

$$\left(\int_X D^q\left(\left(M_n^{\mathcal{F}R}(f,x,s), f(x,s)\right)\right) P(ds)\right)^{\frac{1}{q}}$$
$$\leq \frac{\sum_{k_1=0}^{n} \dots \sum_{k_d=0}^{n} E\left(\frac{T_1 n(x_1 - x_{k_1 1})}{b_1 - a_1}, \dots, \frac{T_d n(x_d - x_{k_d d})}{b_d - a_d}\right)}{W}$$
$$\cdot \left(\int_X D^q\left(f\left(x_{k_1 1}, \dots, x_{k_d d}, s\right), f(x,s)\right) P(ds)\right)^{\frac{1}{q}} \tag{11.325}$$
$$\leq \frac{2^d E^*}{E\left(\frac{T_1}{2}, \dots, \frac{T_d}{2}\right)} \Omega_1^{(\mathcal{F})}\left(f, \frac{\sum_{i=1}^{d}(b_i - a_i)}{n}\right)_{L^q}.$$

We have proved the following approximation result.

Theorem 11.31. *Let* $(X, \mathcal{B}, P)$ *probability space,* $f \in C_{\mathcal{F}R}^{U_q}\left(\prod_{i=1}^{d}[a_i, b_i]\right)$, $1 \leq q < \infty$. *Then*

$$\left\|\left(\int_X D^q\left(\left(M_n^{\mathcal{F}R}(f,x,s), f(x,s)\right)\right) P(ds)\right)^{\frac{1}{q}}\right\|_{\infty,x} \tag{11.326}$$
$$\leq \frac{2^d E^*}{E\left(\frac{T_1}{2}, \dots, \frac{T_d}{2}\right)} \Omega_1^{(\mathcal{F})}\left(f, \frac{\sum_{i=1}^{d}(b_i - a_i)}{n}\right)_{L^q},$$

where $x \in \prod_{i=1}^{d}[a_i, b_i]$, $\forall\, n \in \mathbb{N}$.

Bibliography

1. G.A. Anastassiou, *Rate of convergence of some neural network operators to the unit-univariate case*, J. Math. Anal. Appl. 212 (1997), 237-262.
2. G.A. Anastassiou, *Rate of convergence of some multivariate neural network operators to the unit*, Comput. Math. 40 (2000), 1-19.
3. G.A. Anastassiou, *Quantitative Approximation*, Chapmann & Hall/CRC, Boca Raton, New York, 2001.
4. G.A. Anastassiou, *Fuzzy approximation by fuzzy convolution type operators*, Comput. Math. 48 (2004), 1369-1386.
5. G.A. Anastassiou, *Higher order fuzzy approximation by fuzzy wavelet type and neural network operators*, Comput. Math. 48 (2004), 1387-1401.
6. G.A. Anastassiou, *Higher order fuzzy Korovkin theory via inequalities*, Commun. Appl. Anal. 10 (2) (2006), 359-392.
7. G.A. Anastassiou, *Fuzzy Korovkin theorems and inequalities*, J. Fuzzy Math. 15 (1) (2007), 169-205.
8. G.A. Anastassiou, *Fractional Korovkin theory*, Chaos, Solitons Fractals 42 (4) (2009), 2080-2094.
9. G.A. Anastassiou, *On right fractional calculus*, Chaos, Solitons Fractals 42 (2009), 365-376.
10. G.A. Anastassiou, *Fractional Differentiation Inequalities*, Springer, New York, 2009.
11. G.A. Anastassiou, *Fuzzy Mathematics: Approximation Theory*, Springer, Heidelberg, New York, 2010.
12. G.A. Anastassiou, *Intelligent Systems: Approximation by Artificial Neural Networks*, Springer, Heidelberg, 2011.
13. G.A. Anastassiou, *Fuzzy fractional calculus and Ostrowski inequality*, J. Fuzzy Math. 19 (3) (2011), 577-590.
14. G.A. Anastassiou, *Multivariate hyperbolic tangent neural network approximation*, Comput. Math. 61 (2011), 809-821.
15. G.A. Anastassiou, *Advanced Inequalities*, World Scientific Publishing Corporation, Singapore, 2011.
16. G.A. Anastassiou, *Univariate hyperbolic tangent neural network approximation*, Math. Comput. Model. 53 (2011), 1111-1132.
17. G.A. Anastassiou, *Multivariate sigmoidal neural network approximation*, Neural Netw. 24 (2011), 378-386.
18. G.A. Anastassiou, *Higher order multivariate fuzzy approximation by multivariate fuzzy wavelet type and neural network operators*, J. Fuzzy Math. 19 (3) (2011), 601-618.

19. G.A. Anastassiou, *Univariate sigmoidal neural network approximation*, J. Comput. Anal. Appl. 14 (4) (2012), 659-690.
20. G.A. Anastassiou, *High degree multivariate fuzzy approximation by quasi-interpolation neural network operators*, Discontinuity, Nonlinearity Complexity 2 (2) (2013), 125-146.
21. G.A. Anastassiou, *Approximation by neural network iterates*, in *Advances in Applied Mathematics and Approximation Theory: Contributions from AMAT 2012*, pp. 1-20, Editors: G. Anastassiou and O. Duman, Springer, New York, 2013.
22. G.A. Anastassiou, *Multivariate fuzzy-random normalized neural network approximation operators*, Ann. Fuzzy Math. Inform. 6 (1) (2013), 191-212.
23. G.A. Anastassiou, *Multivariate fuzzy-random quasi-interpolation neural network approximation operators*, J. Fuzzy Math. 22 (1) (2014), 167-184.
24. G.A. Anastassiou, *Approximation by interpolating neural network operators*, Neural, Parallel, Scientific Comput., accepted 2014.
25. G.A. Anastassiou, S. Gal, *On a fuzzy trigonometric approximation theorem of Weierstrass-type*, J. Fuzzy Math. 9 (3) (2001), 701-708.
26. P.L. Butzer, R.J. Nessel, *Fourier Analysis and Approximation*, Pure and Applied Mathematics 40, Academic Press, New York, London, 1971.
27. F.L. Cao, Y.Q. Zhang, *Interpolation and approximation by neural networks in metric spaces.* (Chinese) Acta Math. Sinica (Chin. Ser.) 51 (1) (2008), 91-98.
28. D. Costarelli, *Interpolation by neural network operators activated by ramp functions*, J. Math. Anal. Appl. 419 (1) (2014), 574-582.
29. K. Diethelm, *The Analysis of Fractional Differential Equations*, Lecture Notes in Mathematics 2004, Springer-Verlag, Berlin, Heidelberg, 2010.
30. R.M. Dudley, *Real Analysis and Probability*, Wadsworth & Brooks/Cole Mathematics Series, Pacific Grove, CA, 1989.
31. A.M.A. El-Sayed, M. Gaber, *On the finite Caputo and finite Riesz derivatives*, Electron. J. Theoret. Phys. 3 (12) (2006), 81-95.
32. G.S. Frederico, D.F.M. Torres, *Fractional Optimal Control in the sense of Caputo and the fractional Noether's theorem*, Internat. Mathe. Forum 3 (10) (2008), 479-493.
33. S. Gal, *Approximation theory in fuzzy setting*, Chapter 13 in *Handbook of Analytic-Computational Methods in Applied Mathematics*, 617-666, edited by G. Anastassiou, Chapman & Hall/CRC, Boca Raton, New York, 2000.
34. R. Goetschel Jr., W. Voxman, *Elementary fuzzy calculus*, Fuzzy Sets Syst. 18 (1986), 31-43.
35. O. Kaleva, *Fuzzy differential equations*, Fuzzy Sets Syst. 24 (1987), 301-317.
36. Y.K. Kim, B.M. Ghil, *Integrals of fuzzy-number-valued functions*, Fuzzy Sets Syst. 86 (1997), 213-222.
37. B. Lenze, *Local behaviour of neural network operators-approximation and interpolation*, Analysis 13 (4) (1993), 377-387.
38. B. Lenze, *One-sided approximation and interpolation operators generating hyperbolic sigma-pi neural networks.* in *Multivariate Approximation and Splines* (Mannheim, 1996), 99-112, Internat. Ser. Numer. Math., 125, Birkhäuser, Basel, 1997.
39. A. Pinkus, *Approximation theory of the MLP model in neural networks*, Acta Numer. 8 (1999), 143-195.
40. S.G. Samko, A.A. Kilbas, O.I. Marichev, *Fractional Integrals and Derivatives, The-*

ory and Applications (Gordon and Breach, Amsterdam, 1993) [English translation from the Russian, Integrals and Derivatives of Fractional Order and Some of Their Applications (Nauka i Tekhnika, Minsk, 1987)].

41. C. Wu, Z. Gong, *On Henstock integrals of interval-valued functions and fuzzy valued functions*, Fuzzy Sets Syst. 115 (3) (2000) 377-391.
42. C. Wu, Z. Gong, *On Henstock integral of fuzzy-number-valued functions (I)*, Fuzzy Sets Syst. 120 (3) (2001), 523-532.
43. C. Wu, M. Ma, *On embedding problem of fuzzy number spaces: Part 1*, Fuzzy Sets Syst. 44 (1991), 33-38.

Chapter 12

Approximation and Functional Analysis over Time Scales

Here we start by proving the Riesz representation theorem for positive linear functionals on the space of continuous functions over a time scale. Then we prove further properties for the related Riemann-Stieltjes integral on time scales and we prove the related Hölder's inequality. Next we prove the Hölder's inequality for general positive linear functionals on time scales. We introduce basic concepts of Approximation theory on time scales and we discuss some limitations of the modulus of continuity there. Next we prove the famous Korovkin theorem on time scales, regarding the approximation of unit operator by sequences of positive linear operators on the space of continuity functions defined on a compact interval of a time scale. Then we produce several Shisha-Mond type inequalities related to Korovkin's theorem, putting the convergence of positive linear operators and positive linear functionals in a quantitative form and giving rates of convergence, all operating on Lipschitz functions on a time scale. At the end we present an example of a concrete and genuine positive linear operator on time scales and we give its approximation and interpolation properties over continuous functions. It follows [2].

12.1 Introduction

Denote by $C([a,b])$, the space of continuous real valued functions on $[a,b] \subset \mathbb{R}$.

Definition 12.1. Let L be a linear operator mapping $C([a,b])$ into itself. L is called positive iff whenever $f \geq g$; $f, g \in C([a,b])$, we have that $L(f) \geq L(g)$.

We mention the famous Korovkin's Theorem

Theorem 12.1. *(Korovkin [11] (1960), p. 14). Let $[a,b]$ be a compact interval in $\mathbb{R}$ and $(L_n)_{n\in\mathbb{N}}$ be a sequence of positive linear operators L_n mapping $C([a,b])$ into itself. Suppose that $(L_n f)$ converges uniformly to f for the three test functions $f = 1, x, x^2$. Then $(L_n f)$ converges uniformly to f on $[a,b]$ for all functions $f \in C([a,b])$.*

We need

Definition 12.2. Let $f \in C([a,b])$ and $0 \leq h \leq b-a$. Call

$$\omega_1(f,h) := \sup_{\substack{\text{all } x,y \\ |x-y| \leq h}} |f(x) - f(y)|, \tag{12.1}$$

the first modulus of continuity of f at h.

We also mention the famous theorem

Theorem 12.2. *(Shisha and Mond [13] (1968)). Let $[a,b] \subset \mathbb{R}$ be a compact interval. Let $\{L_n\}_{n \in \mathbb{N}}$ be a sequence of positive linear operators acting on $C([a,b])$ into itself. For $n = 1, 2, ...,$ suppose $L_n(1)$ is bounded. Let $f \in C([a,b])$. Then for $n = 1, 2, ...,$ we have*

$$\|L_n f - f\|_\infty \leq \|f\|_\infty \cdot \|L_n 1 - 1\|_\infty + \|L_n(1) + 1\|_\infty \omega_1(f, \mu_n), \tag{12.2}$$

where

$$\mu_n := \left\| L_n\left((t-x)^2\right)(x) \right\|_\infty^{\frac{1}{2}},$$

and $\|\cdot\|_\infty$ stands for the sup-norm over $[a,b]$. In particular, if $L_n(1) = 1$ then (12.2) reduces to

$$\|L_n(f) - f\|_\infty \leq 2\omega_1(f, \mu_n). \tag{12.3}$$

Note. (i) In forming μ_n^2, x is kept fixed, however t forms the functions t, t^2 on which L_n acts.

(ii) One can easily see, for $n = 1, 2, ...$

$$\mu_n^2 \leq \left\| \left(L_n\left(t^2\right)\right)(x) - x^2 \right\|_\infty + 2c \left\| (L_n(t))(x) - x \right\|_\infty + c^2 \left\| (L_n(1))(x) - 1 \right\|_\infty, \tag{12.4}$$

where $c := \max(|a|, |b|)$.

So if the assumptions of Korovkin's Theorem 12.1 are fulfilled then $\mu_n \to 0$, therefore $\omega_1(f, \mu_n) \to 0$, as $n \to +\infty$, and we obtain from (12.2) that $\|L_n f - f\|_\infty \to 0$, as $n \to +\infty$, which is Korovkin's conclusion!!! I.e., Korovkin's result has been recast in a quantitative form.

Next we mention from [1].

We need

Definition 12.3. Let $B : \mathbb{R} \to \mathbb{R}_+$, be a bell-shaped function of compact support $[-T, T]$, $T > 0$. We assume it is even, non-decreasing for $x < 0$ and non-increasing for $x \geq 0$. Suppose also that $B(0) =: B^* > 0$ is the global maximum of B. The function B may have jump discontinuities and it is measurable. Assume further that $B(\pm T) = 0$.

An example for B can be the hat function

$$\beta(x) := \begin{cases} 1+x, & -1 \leq x \leq 0, \\ 1-x, & 0 < x \leq 1, \\ 0, & \text{elsewhere,} \end{cases} \tag{12.5}$$

etc.

Definition 12.4. Let $f : [a,b] \to \mathbb{R}$, $a, b \in \mathbb{R}$, $a < b$, a bounded and measurable function, $n \in \mathbb{N}$, $h := \frac{b-a}{n}$, $x_k := a + kh$, $k = 0, 1, ..., n$, $x \in [a,b]$.

We define the interpolation neural network operator

$$H_n(f,x) := \frac{\sum_{k=0}^{n} f(x_k) B\left(\frac{Tn(x-x_k)}{b-a}\right)}{\sum_{k=0}^{n} B\left(\frac{Tn(x-x_k)}{b-a}\right)}. \tag{12.6}$$

This is a positive linear operator.

We state the interpolation result.

Theorem 12.3. *([1]) Let $f : [a,b] \to \mathbb{R}$ be a bounded and measurable function. Then*

$$H_n(f, x_i) = f(x_i), \quad i = 0, 1, ..., n, \tag{12.7}$$

where $x_i := a + ih$, $h := \frac{b-a}{n}$, $n \in \mathbb{N}$.

We state the related approximation result at Jackson speed of convergence $\frac{1}{n}$.

Theorem 12.4. *([1]) Let $f \in C([a,b])$. Then*

$$\|H_n(f) - f\|_\infty \leq \frac{2B^*}{B\left(\frac{T}{2}\right)} \omega_1\left(f, \frac{b-a}{n}\right), \quad \forall\, n \in \mathbb{N}. \tag{12.8}$$

The above results motivated this chapter.

In this chapter we prove similar to the above approximation theorems on time scales and we expand around, also treating related Functional Analysis topics.

To our knowledge this is a new study about classical approximation by positive linear operators on time scales.

12.2 Time Scales Basics (See [5])

Since a time scale $\mathbb{T}$ is a closed subset of the real numbers $\mathbb{R}$, it is a complete metric space with the metric (distance)

$$d(t,s) = |t-s|, \quad \text{for } t, s \in \mathbb{T}. \tag{12.9}$$

Consequently, according to basic theory of general metric spaces we have for $\mathbb{T}$ fundamental concepts such as open ball (intervals), neighborhoods of points, open sets, closed sets, compact sets and so on. In particular, for a given $\delta > 0$, the δ-neighborhood $U_\delta(t)$ of a given point $t \in \mathbb{T}$ is the set of all points $s \in \mathbb{T}$ such that $d(t, s) < \delta$. By a neighborhood of a point $t \in \mathbb{T}$ we mean an arbitrary set in $\mathbb{T}$ containing a δ-neighborhood of the point t. Also we have for functions $f : \mathbb{T} \to \mathbb{R}$ the concepts of limit, continuity, and the properties of continuous functions on general complete metric spaces (note that, in particular, any function $f : \mathbb{Z} \to \mathbb{R}$ is continuous at each point of $\mathbb{Z}$).

Let $\mathbb{T}$ be a time scale. We define the forward jump operator $\sigma : \mathbb{T} \to \mathbb{T}$ by

$$\sigma(t) = \inf\{s \in \mathbb{T} : s > t\} \quad \text{for } t \in \mathbb{T},$$

while the backward jump operator $\rho : \mathbb{T} \to \mathbb{T}$ is defined by

$$\rho(t) = \sup\{s \in \mathbb{T} : s < t\} \quad \text{for } t \in \mathbb{T}.$$

In this definition we set in addition $\sigma(\max \mathbb{T}) = \max \mathbb{T}$ if there exists a finite $\max \mathbb{T}$, and $\rho(\min \mathbb{T}) = \min \mathbb{T}$ if there exists a finite $\min \mathbb{T}$.

Obviously both $\sigma(t)$ and $\rho(t)$ are in $\mathbb{T}$ when $t \in \mathbb{T}$. This is because $\mathbb{T}$ is a closed subset of $\mathbb{R}$.

Let $t \in \mathbb{T}$. If $\sigma(t) > t$, we say that t is right-scattered, while if $\rho(t) < t$, we say that t is left-scattered. Also, if $t < \max \mathbb{T}$ and $\sigma(t) = t$, then t is called right-dense, while if $t > \min \mathbb{T}$ and $\rho(t) = t$, then t is called left-dense. Points that are right-dense and left-dense are called dense and points that are right-scattered and left-scattered at the same time are called isolated.

If $\mathbb{T} = \mathbb{R}$, then $\sigma(t) = \rho(t) = t$. If $\mathbb{T} = h\mathbb{Z}$, then $\sigma(t) = t + h$ and $\rho(t) = t - h$. But if $\mathbb{T} = q^{\mathbb{N}_0}$ $(q > 1)$, $\mathbb{N}_0 = \mathbb{N} \cup \{0\}$, then $\sigma(t) = qt$ and $\rho(1) = 1$, $\rho(t) = q^{-1}t$ for $t > 1$.

Let $\mathbb{T}^k$ denote Hilger's truncated set consisting of $\mathbb{T}$ except for a possible left-scattered maximal point. Now we consider a function $f : \mathbb{T} \to \mathbb{R}$ and define the so-called delta (or Hilger) derivative of f at a point $t \in \mathbb{T}^k$.

Assume $f : \mathbb{T} \to \mathbb{R}$ is a function and $t \in \mathbb{T}^k$. Then we define $f^\Delta(t)$ to be the number (provided it exists) with the property that given any $\varepsilon > 0$, there is a neighborhood U (in $\mathbb{T}$) of t such that

$$\left|f(\sigma(t)) - f(s) - f^\Delta(t)[\sigma(t) - s]\right| \le \varepsilon |\sigma(t) - s| \quad \text{for all } s \in U. \tag{12.10}$$

We call $f^\Delta(t)$ the delta (or Hilger) derivative of f at t.

If $t \in \mathbb{T} \backslash \mathbb{T}^k$, then $f^\Delta(t)$ is not uniquely defined, since for such a point t, small neighborhoods U of t consist only of t, and besides, we have $\sigma(t) = t$. Therefore (12.10) holds for an arbitrary number $f^\Delta(t)$.

Note that for the calculation of the delta derivative it is convenient to use its definition in the limit form

$$f^\Delta(t) = \lim_{s \to t} \frac{f(\sigma(t)) - f(s)}{\sigma(t) - s}, \tag{12.11}$$

where the limit is taken in $\mathbb{T}$ with the metric (12.9) and in the limit we suppose $s \neq \sigma(t)$.

Using (12.11) we get the following. If $\mathbb{T} = \mathbb{R}$, then $f^{\Delta}(t) = f'(t)$, the ordinary derivative of f at t. If $\mathbb{T} = h\mathbb{Z}$ $(h > 0)$, then

$$f^{\Delta}(t) = \frac{f(t+h) - f(t)}{h}.$$

In particular, in the case $\mathbb{T} = \mathbb{Z}$, we have

$$f^{\Delta}(t) = f(t+1) - f(t).$$

If $\mathbb{T} = q^{\mathbb{N}_0}$ $(q > 1)$, then

$$f^{\Delta}(t) = \frac{f(qt) - f(t)}{(q-1)t}.$$

For the Riemann-Stieltjes integral on time scales and properties we follow [10]. We write "$f \in \mathcal{R}(\alpha)$ on $[a,b]_{\mathbb{T}}$", if f is Riemann-Stieltjes integrable with respect to a function α on $[a,b]_{\mathbb{T}}$.

We mention

Proposition 12.1. *([10]) If $f : [a,b]_{\mathbb{T}} \to \mathbb{R}$ is continuous on $[a,b]_{\mathbb{T}}$ and if $\alpha : [a,b]_{\mathbb{T}} \to \mathbb{R}$ is of bounded variation on $[a,b]_{\mathbb{T}}$, then $f \in \mathcal{R}(\alpha)$ on $[a,b]_{\mathbb{T}}$, i.e. it exists the Riemann-Stieltjes (R-S) integral on $[a,b]_{\mathbb{T}}$, denoted by $\int_a^b f(t)\Delta\alpha(t)$ or by $\int_a^b f\Delta\alpha$.*

The description of (R-S) integral and properties on time scales parallels to the one on $\mathbb{R}$ of the ordinary case.

Example 12.1. For many interesting examples of time scales see [5].

We give here some important related examples:

(i) Let $0 < r < 1$ fixed. The set $K := \{r^n : n \in \mathbb{N}\} \cup \{0\}$ is a time scale.

(ii) The set $\theta := \left\{\frac{1}{n} : n \in \mathbb{N}\right\} \cup \{0\}$ is a time scale.

Call $T_m := m + \theta$, where $m \in \mathbb{Z}$, then T_m is a time scale.

Set $W := \cup_{m\in\mathbb{Z}} T_m$, then W is a time scale.

Call $B_m := m - \theta$, $m \in \mathbb{Z}$, which is a time scale.

Set $\Lambda := \cup_{m\in\mathbb{Z}} B_m$, then Λ is a time scale. Also $W \cup \Lambda$ is another time scale.

Notice the above sets $K, \theta, T_m, W, B_m, \Lambda, W \cup \Lambda$ contain all of their limit points, where near them the elements of these sets can be arbitrarily close to each other.

We need

Definition 12.5. Let $\mathbb{T}$ be a time scale and $f : \mathbb{T} \to \mathbb{R}$. If

$$|f(x) - f(y)| \leq M |x - y|^{\beta}, \ \ 0 < \beta \leq 1, \tag{12.12}$$

$M > 0$, $\forall\ x, y \in \mathbb{T}$, we call f a Lipschitz function of order β and we denote it as $f \in Lip(\beta)$.

We make

Remark 12.1. Let $f : [a,b]_{\mathbb{T}} \to \mathbb{R}$ be continuous on $[a,b]_{\mathbb{T}}$, with f^{Δ} existing and bounded on $[a,b)_{\mathbb{T}}$, i.e. $\left|f^{\Delta}(t)\right| \leq A$, $\forall\, t \in [a,b)_{\mathbb{T}}$, where $A > 0$. Let $a \leq c < d \leq b$. By the mean-value theorem on time scales ([8]), we have

$$f^{\Delta}(\tau) \leq \frac{f(d) - f(c)}{d-c} \leq f^{\Delta}(\xi), \tag{12.13}$$

where $\tau, \xi \in [c,d)_{\mathbb{T}}$.

Then it follows from (12.13) that

$$|f(d) - f(c)| \leq A(d-c), \tag{12.14}$$

thus $f \in Lip(1)$ on $[a,b]_{\mathbb{T}}$.

12.3 More on Riemann-Stieltjes Integral on Time Scales

Denote by $C([a,b]_{\mathbb{T}})$ the Banach space of all functions $f : [a,b]_{\mathbb{T}} \to \mathbb{R}$ which are continuous (in the metric of the time scale) on $[a,\rho(b)]_{\mathbb{T}}$ and such that $f(b) = f(\rho(b))$, equipped with the norm

$$\|f\|_{\infty} = \max|f(t)|, \quad t \in [a,b]_{\mathbb{T}},$$

where ρ is the backward jump operator in $\mathbb{T}$.

By $B([a,b]_{\mathbb{T}})$ we denote the Banach space of all bounded real-valued functions on $[a,b]_{\mathbb{T}}$ with norm defined by $\|f\|_{\infty} = \sup|f(t)|$, $t \in [a,b]_{\mathbb{T}}$.

We need

Fundamental Lemma A ([6, p. 120]) For every $\delta > 0$ there exists at least one partition $P = \{a = t_0, t_1, ..., t_n = b\}$ of $[a,b]_{\mathbb{T}}$ $(t_0 < t_1 < ... < t_n)$ such that for each $k \in \{1,2,...,n\}$ either $t_k - t_{k-1} \leq \delta$ or $t_k - t_{k-1} > \delta$ and $\rho(t_k) = t_{k-1}$.

For given $\delta > 0$ we denote by $\mathcal{P}_{\delta}([a,b]_{\mathbb{T}})$ the set of all partitions $P = \{t_0, t_1, ..., t_n\}$ of $[a,b]_{\mathbb{T}}$ that possess the property of Fundamental Lemma A.

We present the Riesz representation theorem for positive linear functionals on $C([a,b]_{\mathbb{T}})$.

Theorem 12.5. *Every positive linear functional F on $C([a,b]_{\mathbb{T}})$ can be represented by a Riemann-Stieltjes Δ-integral in the form*

$$F(f) = \int_a^b f(t)\,\Delta\alpha(t), \tag{12.15}$$

$\forall\, f \in C([a,b]_{\mathbb{T}})$, *where α is an increasing function on $[a,b]_{\mathbb{T}}$ with $\|F\| = F(1) = \alpha(b)$, and $\alpha(a) = 0$.*

Proof. Let F be a positive linear functional on $C([a,b]_{\mathbb{T}})$, which is a subspace of $B([a,b]_{\mathbb{T}})$. By the extension theorem on positive linear functional ([12]), there exists a positive linear functional on $B([a,b]_{\mathbb{T}})$, denoted by $\overline{F}$ such that

$$F = \overline{F}|_{C([a,b]_{\mathbb{T}})}. \tag{12.16}$$

We consider the functions χ_t defined for each fixed $t \in [a,b]_{\mathbb{T}}$ by

$$\chi_t(\xi) := \begin{cases} 1, & \text{for } \xi \in [a,t)_{\mathbb{T}}, \\ 0, & \text{for } \xi \in [t,b]_{\mathbb{T}}, \end{cases} \tag{12.17}$$

(for $t \in [a,b)_{\mathbb{T}}$), and

$$\chi_b(\xi) = 1, \text{ for all } \xi \in [a,b]_{\mathbb{T}}.$$

Clearly here $\chi_t \in B([a,b]_{\mathbb{T}})$.

When $t_1 \leq t_2$, $t_1, t_2 \in [a,b)_{\mathbb{T}}$, we get $\chi_{t_1}(\xi) \leq \chi_{t_2}(\xi)$, $\forall\, \xi \in [a,b]_{\mathbb{T}}$, that is $\chi_{t_1} \leq \chi_{t_2} \leq \chi_b$. Hence

$$\overline{F}(\chi_{t_1}) \leq \overline{F}(\chi_{t_2}) \leq \overline{F}(\chi_b) = \overline{F}(1) = F(1). \tag{12.18}$$

We define the function

$$\alpha(t) = \overline{F}(\chi_t), \quad \forall\, t \in [a,b]_{\mathbb{T}}.$$

Notice also $\alpha(a) = \overline{F}(\chi_a) = \overline{F}(0) = 0$. We have that $\alpha(t_1) \leq \alpha(t_2)$, hence α is increasing on $[a,b]_{\mathbb{T}}$, thus α is of bounded variation, with total variation $\alpha(b) = F(1)$.

Next we continue as in [10].

Let $f \in C([a,b]_{\mathbb{T}})$, which is uniformly continuous and which means for any $\varepsilon > 0$ there exists a $\delta > 0$ such that $t', t'' \in [a,b]_{\mathbb{T}}$, $|t' - t''| \leq \delta$ implies $|f(t') - f(t'')| < \varepsilon$.

For every partition $P = \{t_0, t_1, ..., t_n\}$ belonging to $\mathcal{P}_\delta([a,b]_{\mathbb{T}})$ we consider the step function $f^{(\varepsilon)}$ defined on $[a,b]_{\mathbb{T}}$ by

$$f^{(\varepsilon)}(t) := f(t_{k-1}), \text{ if } t \in [t_{k-1}, t_k)_{\mathbb{T}}, \quad k = 1, ..., n, \tag{12.19}$$

and

$$f^{(\varepsilon)}(b) := f(t_n) = f(b), \tag{12.20}$$

which can be written in the form

$$f^{(\varepsilon)}(t) = \sum_{k=1}^{n} f(t_{k-1}) \left[\chi_{t_k}(t) - \chi_{t_{k-1}}(t)\right], \quad \text{for all } t \in [a,b]_{\mathbb{T}}. \tag{12.21}$$

We have that

$$\left|f^{(\varepsilon)}(t) - f(t)\right| < \varepsilon, \quad \forall\, t \in [a,b]_{\mathbb{T}}. \tag{12.22}$$

We see that as follows: let $t \in [a,b]_{\mathbb{T}}$, then either $t \in [t_{k-1}, t_k)_{\mathbb{T}}$ for some $k \in \{1, ..., n\}$ or $t = b$. Consider first the case $t \in [t_{k-1}, t_k)_{\mathbb{T}}$.

If $t_k - t_{k-1} \leq \delta$, then

$$\left|f^{(\varepsilon)}(t) - f(t)\right| = |f(t_{k-1}) - f(t)| < \varepsilon.$$

If $t_k - t_{k-1} > \delta$, then $t_{k-1} = \rho(t_k)$ and hence $[t_{k-1}, t_k)_{\mathbb{T}} = \{t_{k-1}\}$ is a singleton, and thus

$$\left|f^{(\varepsilon)}(t) - f(t)\right| = \left|f^{(\varepsilon)}(t_{k-1}) - f(t_{k-1})\right| = |f(t_{k-1}) - f(t_{k-1})| = 0. \tag{12.23}$$

In the case $t = b$, we get

$$\left|f^{(\varepsilon)}(t) - f(t)\right| = |f(b) - f(b)| = 0.$$

Therefore (12.22) is true.

Consequently it holds

$$\left\|f^{(\varepsilon)} - f\right\|_{\infty} < \varepsilon. \tag{12.24}$$

By linearity of $\overline{F}$ we find

$$\overline{F}\left(f^{(\varepsilon)}\right) = \sum_{k=1}^{n} f(t_{k-1})\left[\overline{F}(\chi_{t_k}) - \overline{F}\left(\chi_{t_{k-1}}\right)\right] = \sum_{k=1}^{n} f(t_{k-1})\left[\alpha(t_k) - \alpha(t_{k-1})\right]. \tag{12.25}$$

Therefore $\overline{F}\left(f^{(\varepsilon)}\right)$ is a Riemann-Stieltjes Δ-sum of f with respect to α, corresponding to the partition P.

Consequently it holds

$$\left|\overline{F}\left(f^{(\varepsilon)}\right) - \int_a^b f(t)\,\Delta\alpha(t)\right| < \varepsilon \tag{12.26}$$

for a sufficiently small $\delta > 0$.

Also we find

$$\left|\overline{F}\left(f^{(\varepsilon)}\right) - F(f)\right| = \left|\overline{F}\left(f^{(\varepsilon)}\right) - \overline{F}(f)\right|$$

$$= \left|\overline{F}\left(f^{(\varepsilon)} - f\right)\right| \le \|\overline{F}\| \left\|f^{(\varepsilon)} - f\right\|_{\infty} \le \|\overline{F}\|\,\varepsilon =: (*). \tag{12.27}$$

Let $g \in B([a,b]_{\mathbb{T}})$, we have that $|g| \le \|g\|_{\infty} < \infty$. Hence $-\|g\|_{\infty} \le g \le \|g\|_{\infty}$ and $-\|g\|_{\infty}\overline{F}(1) \le \overline{F}(g) \le \|g\|_{\infty}\overline{F}(1)$, i.e.

$$\left|\overline{F}(g)\right| \le \overline{F}(1)\|g\|_{\infty} = F(1)\|g\|_{\infty}, \tag{12.28}$$

so that $\overline{F}$ is a real valued bounded linear functional on $B([a,b]_{\mathbb{T}})$ with $\left\|\overline{F}\right\|_{\infty} \le F(1) < \infty$.

Similarly F is a real valued bounded linear functional on $C([a,b]_{\mathbb{T}})$, with $\|F\|_{\infty} \le F(1)$.

Furthermore it holds for $g(t) = 1$, $\forall\, t \in [a,b]_{\mathbb{T}}$, that

$$F(1) = |F(1)| \le \|F\|_{\infty} \cdot 1 = \|F\|_{\infty},$$

along with

$$F(1) = \left|\overline{F}(1)\right| \le \left\|\overline{F}\right\|_{\infty} \cdot 1 = \left\|\overline{F}\right\|_{\infty}.$$

We conclude that

$$F(1) = \|F\|_{\infty} = \left\|\overline{F}\right\|_{\infty} = \alpha(b). \tag{12.29}$$

So that

$$(*) = \alpha(b)\varepsilon.$$

I.e.

$$\left|\overline{F}\left(f^{(\varepsilon)}\right) - F(f)\right| \leq \alpha(b) \cdot \varepsilon. \tag{12.30}$$

Finally we derive

$$\left|F(f) - \int_a^b f(t)\,\Delta\alpha(t)\right| \leq \left|F(f) - \overline{F}\left(f^{(\varepsilon)}\right)\right| + \left|\overline{F}\left(f^{(\varepsilon)}\right) - \int_a^b f(t)\,\Delta\alpha(t)\right| \tag{12.31}$$

$$\leq \alpha(b) \cdot \varepsilon + \varepsilon = \varepsilon(\alpha(b) + 1),$$

which implies (12.15), because $\varepsilon > 0$ is arbitrary. The proof of the theorem now is complete. □

We make

Remark 12.2. Here $\alpha : [a,b]_{\mathbb{T}} \to \mathbb{R}$ is assumed to be increasing and let $f \in C([a,b]_{\mathbb{T}})$. Clearly α is of bounded variation and $f, |f| \in \mathcal{R}(\alpha)$ on $[a,b]_{\mathbb{T}}$. We notice that

$$\left|\int_a^b f(t)\,\Delta\alpha(t)\right| \leq \|f\|_{\infty}(\alpha(b) - \alpha(a)), \tag{12.32}$$

and

$$\left|\int_a^b f(t)\,\Delta\alpha(t)\right| \leq \int_a^b |f(t)|\,\Delta\alpha(t). \tag{12.33}$$

Given $f(t) \geq 0$, $\forall\, t \in [a,b]_{\mathbb{T}}$, we get

$$\int_a^b f(t)\,\Delta\alpha(t) \geq 0. \tag{12.34}$$

Let also $g \in C([a,b]_{\mathbb{T}})$, such that $f(t) \geq g(t)$, $\forall\, t \in [a,b]_{\mathbb{T}}$. Hence $f(t) - g(t) \geq 0$, and

$$\int_a^b f(t)\,\Delta\alpha(t) - \int_a^b g(t)\,\Delta\alpha(t) = \int_a^b (f(t) - g(t))\,\Delta\alpha(t) \geq 0.$$

Thus

$$\int_a^b f(t)\,\Delta\alpha(t) \geq \int_a^b g(t)\,\Delta\alpha(t). \tag{12.35}$$

Clearly if $\alpha = 0$ or a constant, then

$$\int_a^b f(t)\,\Delta\alpha(t) = 0. \tag{12.36}$$

Let now α be strictly increasing on $[a,b]_{\mathbb{T}}$ and $f(t) \geq 0$, $\forall\, t \in [a,b]_{\mathbb{T}}$ with $\int_a^b f(t)\,\Delta\alpha(t) = 0$. Then $f(t) = 0$, $\forall\, t \in [a,b]_{\mathbb{T}}$. If $f(t) > 0$, $\forall\, t \in [a,b]_{\mathbb{T}}$, α is increasing on $[a,b]_{\mathbb{T}}$ and $\int_a^b f(t)\,\Delta\alpha(t) = 0$. Then α is either zero, or a constant different than zero.

Next comes Hölder's inequality for (R-S) integrals

Theorem 12.6. *Let $f, g \in C([a,b]_{\mathbb{T}})$, α is increasing on $[a,b]_{\mathbb{T}}$, and $p, q > 1$: $\frac{1}{p}+\frac{1}{q}=1$. Then*

$$\int_a^b |f(t)\, g(t)|\, \Delta\alpha(t) \leq \left(\int_a^b |f(t)|^p \Delta\alpha(t)\right)^{\frac{1}{p}} \left(\int_a^b |g(t)|^q \Delta\alpha(t)\right)^{\frac{1}{q}}. \quad (12.37)$$

Equality holds nontrivially, when $|g(x)| = c\,|f(x)|^{p-1}$, $c > 0$, $\forall\, x \in [a,b]_{\mathbb{T}}$.

Proof. For $\widetilde{\alpha}, \widetilde{\beta} \geq 0$, the basic inequality holds

$$\widetilde{\alpha}^{\frac{1}{p}} \widetilde{\beta}^{\frac{1}{q}} \leq \frac{\widetilde{\alpha}}{p} + \frac{\widetilde{\beta}}{q}, \quad (12.38)$$

with equality when $\widetilde{\alpha} = \widetilde{\beta}$.

Now suppose, without loss of generality, that

$$\int_a^b |f(t)|^p \Delta\alpha(t), \int_a^b |g(t)|^q \Delta\alpha(t) \neq 0.$$

Apply inequality (12.38) for

$$\widetilde{\alpha}(t) = \frac{|f(t)|^p}{\int_a^b |f(\tau)|^p \Delta\alpha(\tau)}, \quad \text{and } \widetilde{\beta}(t) = \frac{|g(t)|^q}{\int_a^b |g(\tau)|^q \Delta\alpha(\tau)}, \quad (12.39)$$

to get

$$\left(\frac{|f(t)|}{\left(\int_a^b |f(\tau)|^p \Delta\alpha(\tau)\right)^{\frac{1}{p}}}\right) \left(\frac{|g(t)|}{\left(\int_a^b |g(\tau)|^q \Delta\alpha(\tau)\right)^{\frac{1}{q}}}\right)$$

$$\leq \frac{|f(t)|^p}{p\int_a^b |f(\tau)|^p \Delta\alpha(\tau)} + \frac{|g(t)|^q}{q\int_a^b |g(\tau)|^q \Delta\alpha(\tau)}. \quad (12.40)$$

Hence by integrating (12.40) it holds

$$\frac{\int_a^b |f(t)|\,|g(t)|\,\Delta\alpha(t)}{\left(\int_a^b |f(\tau)|^p \Delta\alpha(\tau)\right)^{\frac{1}{p}} \left(\int_a^b |g(\tau)|^q \Delta\alpha(\tau)\right)^{\frac{1}{q}}}$$

$$\leq \frac{\int_a^b |f(t)|^p \Delta\alpha(t)}{p\int_a^b |f(\tau)|^p \Delta\alpha(\tau)} + \frac{\int_a^b |g(t)|^q \Delta\alpha(t)}{q\int_a^b |g(\tau)|^q \Delta\alpha(\tau)} = \frac{1}{p} + \frac{1}{q} = 1, \quad (12.41)$$

proving inequality (12.37).

Next we assume that $|g(x)| = c\,|f(x)|^{p-1}$, $c > 0$. Then

$$\int_a^b |f(t)|\,|g(t)|\,\Delta\alpha(t) = c\left(\int_a^b |f(t)|^p \Delta\alpha(t)\right). \quad (12.42)$$

On the other hand we can write

$$\left(\int_a^b |f(t)|^p \Delta\alpha(t)\right)^{\frac{1}{p}} \left(\int_a^b |g(t)|^q \Delta\alpha(t)\right)^{\frac{1}{q}}$$

$$= \left(\int_a^b |f(t)|^p \Delta\alpha(t)\right)^{\frac{1}{p}} c \left(\int_a^b |f(t)|^{q(p-1)} \Delta\alpha(t)\right)^{\frac{1}{q}}$$

$$= c\left(\int_a^b |f(t)|^p \Delta\alpha(t)\right)^{\frac{1}{p}} \left(\int_a^b |f(t)|^p \Delta\alpha(t)\right)^{\frac{1}{q}} = c\left(\int_a^b |f(t)|^p \Delta\alpha(t)\right). \tag{12.43}$$

By (12.42), (12.43) we have proved that (12.37) is attained.

The proof of the theorem now is complete. □

The special case $p = q = 2$ reduces to the Cauchy-Schwarz inequality.

Corollary 12.1. *(to Theorem 12.6) It holds*

$$\int_a^b |f(t)\, g(t)| \Delta\alpha(t) \le \left(\int_a^b f^2(t) \Delta\alpha(t)\right)^{\frac{1}{2}} \left(\int_a^b g^2(t) \Delta\alpha(t)\right)^{\frac{1}{2}}. \tag{12.44}$$

12.4 Approximation Basics on Time Scales

We need

Definition 12.6. Let $f \in B([a,b]_{\mathbb{T}})$. We define

$$\omega_1^{\mathbb{T}}(f,\delta) := \sup_{\substack{x,y\in[a,b]_{\mathbb{T}}:\\ |x-y|\le\delta}} |f(x) - f(y)|, \quad 0 < \delta \le b-a, \tag{12.45}$$

and

$$\omega_1^{\mathbb{T}}(f,\delta) = \omega_1(f, b-a), \quad \text{if } \delta > b-a. \tag{12.46}$$

We call $\omega_1^{\mathbb{T}}(f,\cdot)$ the first modulus of continuity of f.

Theorem 12.7. *It holds*

(i) $\omega_1^{\mathbb{T}}(f,\delta) < \infty$,

(ii) $\omega_1^{\mathbb{T}}(f,\cdot)$ *is non-negative and increasing on* $\mathbb{R}_+$,

(iii)

$$\omega_1^{\mathbb{T}}(f,0) = 0, \tag{12.47}$$

(iv) $\lim_{\delta\downarrow 0}\omega_1^{\mathbb{T}}(f,\delta) = 0$, *iff f is uniformly continuous on* $[a,b]_{\mathbb{T}}$.

Proof. (i), (ii) and (iii) are obvious.

(iv) ($\Rightarrow$) Let $\omega_1^{\mathbb{T}}(f,\delta) \to 0$ as $\delta \downarrow 0$. Then $\forall\, \varepsilon > 0$, $\exists\, \delta > 0$ with $\omega_1^{\mathbb{T}}(f,\delta) \le \varepsilon$. I.e. $\forall\, x, y \in [a,b]_{\mathbb{T}} : |x-y| \le \delta$ we get $|f(x) - f(y)| \le \varepsilon$. That is f is uniformly continuous on $[a,b]_{\mathbb{T}}$.

($\Leftarrow$) Let f be uniformly continuous on $[a,b]_{\mathbb{T}}$. Then $\forall\, \varepsilon > 0, \exists\, \delta > 0$: whenever $|x-y| \le \delta$, $x, y \in [a,b]_{\mathbb{T}}$, it implies $|f(x) - f(y)| \le \varepsilon$. I.e. $\forall\, \varepsilon > 0$, $\exists\, \delta > 0$: $\omega_1^{\mathbb{T}}(f,\delta) \le \varepsilon$. That is $\omega_1^{\mathbb{T}}(f,\delta) \to 0$ as $\delta \downarrow 0$. □

Comment. (i) $\omega_1^{\mathbb{T}}(f,\cdot)$ fails the subadditivity property and other important properties of usual first modulus of continuity defined on a continuous interval $[a,b]$ or $\mathbb{R}$.

(ii) Let f be continuous on $[a,b]_{\mathbb{T}}$ or $f \in C([a,b]_{\mathbb{T}})$. Since f is continuous (in the time scale topology), it is uniformly continuous on the compact set $[a,b]_{\mathbb{T}}$.

We make

Definition 12.7. (i) Denote by $C_u([a,b]_{\mathbb{T}})$ all the continuous functions from $[a,b]_{\mathbb{T}} \to \mathbb{R}$.

Clearly $C_u([a,b]_{\mathbb{T}})$ is a Banach space, and $C([a,b]_{\mathbb{T}}) \subset C_u([a,b]_{\mathbb{T}})$.

(ii) Let $L : C([a,b]_{\mathbb{T}}) \to C([a,b]_{\mathbb{T}})$ or $L : C_u([a,b]_{\mathbb{T}}) \to C_u([a,b]_{\mathbb{T}})$ a linear operator. Let $f, g \in C_u([a,b]_{\mathbb{T}})$ or in $C([a,b]_{\mathbb{T}})$. The operator L is called positive iff whenever $f \ge g$ on $[a,b]_{\mathbb{T}}$, we have that $L(f) \ge L(g)$ on $[a,b]_{\mathbb{T}}$.

We need (see also [7] in another abstract setting.)

Theorem 12.8. *Let T be a positive linear functional from $C_u([a,b]_{\mathbb{T}})$ into $\mathbb{R}$, and $f \in C_u([a,b]_{\mathbb{T}})$ with $f \ge 0$.*

Then the following are equivalent

(i)

$$T(f) = 0 \tag{12.48}$$

(ii)

$$T(fg) = 0,\ \forall\, g \in C_u([a,b]_{\mathbb{T}}), \tag{12.49}$$

(iii)

$$T(f^m) = 0,\ \textit{for some } m \in \mathbb{N}. \tag{12.50}$$

Proof. (i)$\Rightarrow$(ii). Let any $n \in \mathbb{N}$ and enough to take a $g \in C_u([a,b]_{\mathbb{T}})$ with $g \ge 0$. If $\min\{g,n\} = g$, i.e. $g \le n$, clearly it holds then

$$g \le g + \frac{g^2}{2n}, \tag{12.51}$$

which is always true.

If $\min\{g,n\} = n$, i.e. $n \le g$, since $(g-n)^2 \ge 0 \Leftrightarrow g^2 - 2gn + n^2 \ge 0 \Leftrightarrow g^2 + n^2 \ge 2gn \Leftrightarrow$

$$g \le \frac{g^2+n^2}{2n} = \frac{g^2}{2n} + \frac{n}{2} \le \frac{g^2}{2n} + n. \tag{12.52}$$

So we have proved that

$$0 \le g - \min\{g, n\} \le \frac{g^2}{2n}, \quad \forall\, n \in \mathbb{N}. \tag{12.53}$$

Hence

$$0 \le fg - \min\{fg, nf\} \le \frac{fg^2}{2n}, \quad \forall\, n \in \mathbb{N},$$

so that

$$0 \le T(fg) - T(\min\{fg, nf\}) \le \frac{1}{2n} T\left(fg^2\right), \quad \forall\, n \in \mathbb{N}. \tag{12.54}$$

However it holds

$$0 \le T(\min\{fg, fn\}) \le nT(f) = 0, \forall\, n \in \mathbb{N}. \tag{12.55}$$

That is

$$0 \le T(fg) \le \frac{T\left(fg^2\right)}{2n}, \quad \forall\, n \in \mathbb{N}, \tag{12.56}$$

producing

$$T(fg) = 0. \tag{12.57}$$

(ii)⇒(iii). obvious.

(iii)⇒(i). The result is trivial for $m = 1$, so suppose that $m \ge 2$. We proceed by complete induction hypothesis method. If $m = 2$ and $T\left(f^2\right) = 0$, then

$$0 \le T\left((nf-1)^2\right) = T\left(n^2 f^2 - 2nf + 1\right)$$

$$= n^2 T\left(f^2\right) - 2nT(f) + T(1) = T(1) - 2nT(f). \tag{12.58}$$

So that

$$0 \le 2nT(f) \le T(1), \tag{12.59}$$

and

$$0 \le T(f) \le \frac{T(1)}{2n}, \quad \forall\, n \in \mathbb{N}. \tag{12.60}$$

That is $T(f) = 0$.

Next, let $m \ge 3$ such that $T(f^m) = 0$, we want to prove $T(f) = 0$. Assume for all m' with $2 \le m' < m$ that we have true $T\left(f^{m'}\right) = 0$ implies $T(f) = 0$. Let $k = 0$ or $k = 1$ and particular m' as above such that $m + k = 2m'$.

We use the already proved direction (i)⇒(ii).

Since $T(f^m) = 0$ we get

$$T\left(f^{m+k}\right) = T\left(f^m \cdot f^k\right) = 0.$$

I.e.

$$T\left(f^{m+k}\right) = 0. \tag{12.61}$$

Also it holds

$$T\left(\left(f^{m'}\right)^2\right) = T\left(f^{2m'}\right) = T\left(f^{m+k}\right) = 0, \tag{12.62}$$

i.e.

$$T\left(\left(f^{m'}\right)^2\right) = 0. \tag{12.63}$$

By what we proved earlier in this direction we find $T\left(f^{m'}\right) = 0$. The last by complete induction hypothesis implies $T(f) = 0$. We are done. □

We prove Hölder's inequality for positive linear operators over time scales.

Theorem 12.9. *Let L be a positive linear operator from $C_u([a,b]_{\mathbb{T}})$ into itself and let $p, q > 1 : \frac{1}{p} + \frac{1}{q} = 1$.*

Then:

(i)

$$(L(|fg|))(x) \leq ((L(|f|^p))(x))^{\frac{1}{p}} ((L(|g|^q))(x))^{\frac{1}{q}}, \tag{12.64}$$

for any $x \in [a,b]_{\mathbb{T}}$ and any $f, g \in C_u([a,b]_{\mathbb{T}})$, i.e.

$$L(|fg|) \leq (L(|f|^p))^{\frac{1}{p}} (L(|g|^q))^{\frac{1}{q}}, \tag{12.65}$$

which is attained when $|g| = c|f|^{p-1}$, $c > 0$,

and

(ii)

$$\|L(|fg|)\|_\infty \leq (\|L(|f|^p)\|_\infty)^{\frac{1}{p}} (\|L(|g|^q)\|_\infty)^{\frac{1}{q}}. \tag{12.66}$$

Proof. Notice for a particular $x \in [a,b]_{\mathbb{T}}$, the function $(L(\cdot))(x)$ is a positive linear functional on $C_u([a,b]_{\mathbb{T}})$.

For $\widetilde{\alpha}, \widetilde{\beta} \geq 0$, we have

$$\widetilde{\alpha}^{\frac{1}{p}} \widetilde{\beta}^{\frac{1}{q}} \leq \frac{\widetilde{\alpha}}{p} + \frac{\widetilde{\beta}}{q}, \tag{12.67}$$

with equality when $\widetilde{\alpha} = \widetilde{\beta}$.

Now suppose first that

$$(L(|f|^p))(x), (L(|g|^q))(x) \neq 0.$$

Apply inequality (12.67) for

$$\widetilde{\alpha}(t) = \frac{|f(t)|^p}{(L(|f|^p))(x)} \tag{12.68}$$

and

$$\widetilde{\beta}(t) = \frac{|g(t)|^q}{(L(|g|^q))(x)}, \tag{12.69}$$

to get

$$\frac{|f(t)|}{((L(|f|^p))(x))^{\frac{1}{p}}}\frac{|g(t)|}{((L(|g|^q))(x))^{\frac{1}{q}}} \leq \frac{|f(t)|^p}{p(L(|f|^p))(x)} + \frac{|g(t)|^q}{q(L(|g|^q))(x)}, \tag{12.70}$$

$\forall\, t \in [a,b]_{\mathbb{T}}$.

By positivity and linearity of the functional $(L(\cdot))(x)$, where $x \in [a,b]_{\mathbb{T}}$ is fixed and from (12.70) we find

$$\frac{L(|fg|)(x)}{((L(|f|^p))(x))^{\frac{1}{p}}((L(|g|^q))(x))^{\frac{1}{q}}}$$

$$\leq \frac{(L(|f|^p))(x)}{p(L(|f|^p))(x)} + \frac{(L(|g|^q))(x)}{q(L(|g|^q))(x)} = \frac{1}{p} + \frac{1}{q} = 1, \tag{12.71}$$

that is proving (12.64).

Next we assume $|g| = c\,|f|^{p-1}$, $c > 0$.

Then

$$L(|fg|)(x) = c\,(L(|f|^p)(x)). \tag{12.72}$$

Also it holds

$$((L(|f|^p))(x))^{\frac{1}{p}}((L(|g|^q))(x))^{\frac{1}{q}} = ((L(|f|^p))(x))^{\frac{1}{p}}\,c\left(\left(L\left(|f|^{q(p-1)}\right)\right)(x)\right)^{\frac{1}{q}}$$

$$= c\,((L(|f|^p))(x))^{\frac{1}{p}}((L(|f|^p))(x))^{\frac{1}{q}} = c\,((L(|f|^p))(x)), \tag{12.73}$$

proving that (12.65) is attained.

Assume now that

$$(L(|f|^p))(x) = 0, \;\; p > 1 \text{ with } p \notin \mathbb{N}. \tag{12.74}$$

Consider $\lambda := [p] + 1 - p \geq 0$ ($[\cdot]$ is the integral part), then

$$[p] + 1 = \lambda + p > 1. \tag{12.75}$$

We have

$$L\left(|f|^{[p]+1}\right)(x) = L\left(|f|^{\lambda+p}\right)(x) = L\left(|f|^p \cdot |f|^\lambda\right)(x) = 0. \tag{12.76}$$

The last is true by Theorem 12.8 from the direction (i)$\Rightarrow$(ii).

I.e.

$$L\left(|f|^{[p]+1}\right)(x) = 0. \tag{12.77}$$

Again by Theorem 12.8 from the direction (iii)$\Rightarrow$(ii), we derive

$$(L(|fg|))(x) = L(|f|\,|g|)(x) = 0, \tag{12.78}$$

proving inequality (12.64) is valid as equality with both sides equal to zero.

The case of $p \in \mathbb{N} - \{1\}$ is similar and easier.

The proof of the theorem is now complete. $\square$

12.5 Approximation on Time Scales

We give Korovkin's theorem on time scales.

Theorem 12.10. *Let $[a,b]_{\mathbb{T}} \subset \mathbb{T}$, $\mathbb{T}$ is a time scale, and $(L_n)_{n\in\mathbb{N}}$ be a sequence of positive linear operators, L_n is mapping $C_u([a,b]_{\mathbb{T}})$ into itself. Suppose that $(L_n f)$ converges uniformly to f for the three test functions $f = 1, x, x^2$, $x \in [a,b]_{\mathbb{T}}$.*

Then $(L_n f)$ converges uniformly to f on $[a,b]_{\mathbb{T}}$ for any function $f \in C_u([a,b]_{\mathbb{T}})$.

Proof. (Similar to Bauer (1978), [3])

Denote $f(x) = x$ by id, $f(x) = x^2$ by id^2, where $x \in [a,b]_{\mathbb{T}}$. Every $f \in C_u([a,b]_{\mathbb{T}})$ is bounded:

$$|f(x)| \leq \gamma, \ \forall\, x \in [a,b]_{\mathbb{T}}, \tag{12.79}$$

where $\gamma > 0$.

Also since f is uniformly continuous on $[a,b]_{\mathbb{T}}$, we have $\forall\, \varepsilon > 0\ \exists\, \delta > 0$: for all $x, y \in [a,b]_{\mathbb{T}}$,

$$|x-y| \leq \sqrt{\delta} \text{ then } |f(x) - f(y)| \leq \varepsilon, \tag{12.80}$$

$\Leftrightarrow$

$$(x-y)^2 \leq \delta \text{ then } |f(x) - f(y)| \leq \varepsilon. \tag{12.81}$$

If $(x-y)^2 \leq \delta$ then

$$|f(x) - f(y)| \leq \varepsilon + \alpha^*(x-y)^2, \tag{12.82}$$

with $\alpha^* := 2\gamma\delta^{-1}$,

and if $(x-y)^2 > \delta$, i.e. $\frac{(x-y)^2}{\delta} > 1$ we find that

$$|f(x) - f(y)| \leq 2\gamma \leq \frac{2\gamma}{\delta}(x-y)^2 \tag{12.83}$$

$$= \alpha^*(x-y)^2 \leq \varepsilon + \alpha^*(x-y)^2.$$

So in either case we get

$$|f(x) - f(y)| \leq \varepsilon + \alpha^*(x-y)^2. \tag{12.84}$$

Thus for any $y \in [a,b]_{\mathbb{T}}$ we get

$$|f - f(y)| \leq \varepsilon + \alpha^*(id - y)^2. \tag{12.85}$$

Linearity and positivity of the operators L_n then imply

$$|L_n f - f(y) L_n(1)| \leq \varepsilon L_n(1) + \alpha^*\left[L_n\left(id^2\right) - 2y L_n(id) + y^2 L_n(1)\right]. \tag{12.86}$$

Evaluating the above inequality at $x = y$, we get

$$|L_n f - f \cdot L_n 1| \leq \varepsilon(L_n 1 - 1) + \varepsilon + \alpha^*\left(L_n\left(id^2\right) - 2(id) L_n(id) + (id)^2 L_n(1)\right). \tag{12.87}$$

From the assumption on $(L_n(f))$ for the three functions $f = 1, id, id^2$ and from the triangle inequality it follows that $L_n f$ converges to f uniformly, true for any $f \in C_u([a,b]_{\mathbb{T}})$. □

We give the following quantitative approximation theorem on time scales (see also Theorem 12.2).

Theorem 12.11. *Let $[a,b]_{\mathbb{T}} \subset \mathbb{T}$, $\mathbb{T}$ is a time scale, and $(L_n)_{n\in\mathbb{N}}$ be a sequence of positive linear operators, L_n is mapping $C_u([a,b]_{\mathbb{T}})$ into itself. Here we consider $f \in C_u([a,b]_{\mathbb{T}})$ such that*

$$|f(x) - f(y)| \leq M|x-y|, \quad \forall\, x \in [a,b]_{\mathbb{T}}, \tag{12.88}$$

where $M > 0$.

Then

$$\|L_n f - f\|_\infty \leq \|f\|_\infty \|L_n(1) - 1\|_\infty + M\left(\sqrt{\|L_n(1)-1\|_\infty + 1}\right)$$

$$\left(\sqrt{\|(L_n(t^2))(x) - x^2\|_\infty + 2c\|(L_n(t))(x) - x\|_\infty + c^2\|L_n(1)-1\|_\infty}\right), \tag{12.89}$$

where $c = \max(|a|,|b|)$.

If $L_n(1) = 1$ we get

$$\|L_n f - f\|_\infty \leq M\sqrt{\|(L_n(t^2))(x) - x^2\|_\infty + 2c\|(L_n(t))(x) - x\|_\infty}. \tag{12.90}$$

If $L_n(t^k)(x)$ converges uniformly to x^k on $[a,b]_{\mathbb{T}}$, for $k = 0,1,2$, then by (12.89), we get that $L_n f$ converges uniformly to f over $[a,b]_{\mathbb{T}}$ as $n \to +\infty$.

Given that f^Δ exists and is bounded on $[a,b)_{\mathbb{T}}$, the Lipschitz constant M could have been such that $|f^\Delta(t)| \leq M$, $\forall\, t \in [a,b)_{\mathbb{T}}$, see Remark 12.1.

Proof. Let $x \in [a,b]_{\mathbb{T}}$ be momentarily fixed. We notice the following

$$(L_n f)(x) - f(x) = (L_n f)(x) - f(x)(L_n(1))(x) + f(x)(L_n(1))(x) - f(x)$$

$$= (L_n f)(x) - L_n(f(x))(x) + f(x)((L_n(1))(x) - 1)$$

$$= (L_n(f - f(x)))(x) + f(x)((L_n(1))(x) - 1). \tag{12.91}$$

Therefore we obtain

$$|(L_n f)(x) - f(x)|$$

$$\leq |(L_n(f - f(x)))(x)| + |f(x)||(L_n(1))(x) - 1|$$

$$\leq (L_n(|f - f(x)|))(x) + |f(x)||(L_n(1))(x) - 1| =: (*). \tag{12.92}$$

Since we assumed

$$|f(y) - f(x)| \leq M|y - x|, \tag{12.93}$$

$\forall\, x,y \in [a,b]_{\mathbb{T}}$, we can write

$$|f(\cdot) - f(x)| \leq M|id(\cdot) - x|, \tag{12.94}$$

for any fixed $x \in [a,b]_{\mathbb{T}}$.

So that

$$(*) \leq M\left(L_n\left(|id(\cdot)-x|\right)\right)(x)+|f(x)|\left|\left(L_n(1)\right)(x)-1\right| \tag{12.95}$$

$$\overset{(12.64)}{\leq} M\left(\left(L_n(1)\right)(x)\right)^{\frac{1}{2}}\left(\left(L_n\left(\left(id(\cdot)-x\right)^2\right)\right)(x)\right)^{\frac{1}{2}}+|f(x)|\left|\left(L_n(1)\right)(x)-1\right|$$

$$\leq M\left\|L_n(1)\right\|_{\infty}^{\frac{1}{2}}\left\|L_n\left(\left(id(\cdot)-x\right)^2\right)(x)\right\|_{\infty}^{\frac{1}{2}}+\|f\|_{\infty}\left\|L_n(1)-1\right\|_{\infty}. \tag{12.96}$$

We have derived that

$$\left\|\left(L_n f\right)-f\right\|_{\infty}$$

$$\leq \|f\|_{\infty}\left\|L_n(1)-1\right\|_{\infty}+M\sqrt{\left\|L_n(1)\right\|_{\infty}}\sqrt{\left\|\left(L_n\left((t-x)^2\right)\right)(x)\right\|_{\infty}}$$

$$\leq \|f\|_{\infty}\left\|L_n(1)-1\right\|_{\infty}+M\sqrt{\left\|L_n(1)-1\right\|_{\infty}+1}$$

$$\sqrt{\left\|\left(L_n\left(t^2\right)\right)(x)-x^2\right\|_{\infty}+2c\left\|\left(L_n(t)\right)(x)-x\right\|_{\infty}+c^2\left\|L_n(1)-1\right\|_{\infty}}, \tag{12.97}$$

proving the claim. □

When smoothness is present the speed of convergence improves dramatically. We present

Theorem 12.12. *Let $[a,b]_{\mathbb{T}} \subset \mathbb{T}$, $\mathbb{T}$ is a time scale, and $(L_n)_{n\in\mathbb{N}}$ be a sequence of positive linear operators, L_n is mapping $C_u^1([a,b]_{\mathbb{T}})$ (the space of one time continuously differentiable functions on $[a,b]_{\mathbb{T}}$) into $C_u([a,b]_{\mathbb{T}})$. Here we consider $f \in C_u^1([a,b]_{\mathbb{T}})$ such that*

$$\left|f^{\Delta}(x)-f^{\Delta}(y)\right| \leq M|x-y|, \tag{12.98}$$

$\forall\, x,y \in [a,b]_{\mathbb{T}}$, where $M>0$.

Then

$$\left\|L_n(f)-f\right\|_{\infty} \leq \|f\|_{\infty}\left\|L_n(1)-1\right\|_{\infty}$$

$$+\left\|f^{\Delta}\right\|_{\infty}\left\|\left(L_n(t-x)\right)(x)\right\|_{\infty}+M\left\|\left(L_n\left((t-x)^2\right)\right)(x)\right\|_{\infty}. \tag{12.99}$$

We have that

$$\left\|\left(L_n(t-x)\right)(x)\right\|_{\infty}$$

$$\leq\left(\sqrt{\left\|L_n(1)-1\right\|_{\infty}+1}\right)\sqrt{\left\|\left(\left(L_n(t-x)^2\right)\right)(x)\right\|_{\infty}}, \tag{12.100}$$

and

$$\left\|\left(L_n\left((t-x)^2\right)\right)(x)\right\|_{\infty} \leq \left\|\left(L_n\left(t^2\right)\right)(x)-x^2\right\|_{\infty}$$

$$+2c \left\| \left(L_n \left(t \right) \right) \left(x \right) - x \right\|_{\infty} + c^2 \left\| L_n \left(1 \right) - 1 \right\|_{\infty}, \tag{12.101}$$

where $c := \max \left(|a|, |b| \right)$.

Clearly by (12.99), (12.100), (12.101), under the assumptions $\left(L_n \left(t^k \right) \right) \left(x \right) \to x^k$, $k = 0, 1, 2$, *converge uniformly on* $[a, b]_{\mathbb{T}}$, *we obtain* $L_n f \to f$, $f \in C_u^1 \left([a, b]_{\mathbb{T}} \right)$, *converges uniformly, as* $n \to +\infty$.

If $\left(L_n \left(1 \right) \right) \left(x \right) = 1$ *and* $\left(L_n \left(t \right) \right) \left(x \right) = x$, *we get that*

$$\left(L_n \left(t - x \right)^2 \right) \left(x \right) = \left(L_n \left(t^2 \right) \right) \left(x \right) - x^2, \tag{12.102}$$

and then the speed of uniform convergence of $L_n f \to f$ *squares in comparison to Theorem 12.11, under the assumption that* $\left(L_n \left(t^2 \right) \right) \left(x \right) \to x^2$, *converges uniformly on* $[a, b]_{\mathbb{T}}$, *as* $n \to +\infty$.

Proof. For the Delta Δ-integral on time scales and properties we refer to [5].

By fundamental property

$$\int_x^t f^{\Delta} \left(\tau \right) \Delta \tau = f \left(t \right) - f \left(x \right), \tag{12.103}$$

we have

$$f \left(t \right) = f \left(x \right) + \int_x^t f^{\Delta} \left(\tau \right) \Delta \tau, \tag{12.104}$$

and

$$f \left(t \right) = f \left(x \right) + f^{\Delta} \left(x \right) \left(t - x \right) + \int_x^t \left(f^{\Delta} \left(\tau \right) - f^{\Delta} \left(x \right) \right) \Delta \tau. \tag{12.105}$$

Call

$$\mathcal{R} \left(t, x \right) := \int_x^t \left(f^{\Delta} \left(\tau \right) - f^{\Delta} \left(x \right) \right) \Delta \tau. \tag{12.106}$$

We estimate first $\mathcal{R} \left(t, x \right)$.

(i) If $t \geq x$ we get

$$\left| \mathcal{R} \left(t, x \right) \right| = \left| \int_x^t \left(f^{\Delta} \left(\tau \right) - f^{\Delta} \left(x \right) \right) \Delta \tau \right| \leq \int_x^t \left| f^{\Delta} \left(\tau \right) - f^{\Delta} \left(x \right) \right| \Delta \tau$$

$$\leq M \int_x^t \left| \tau - x \right| \Delta \tau \leq M \left(t - x \right)^2. \tag{12.107}$$

(ii) If $t < x$ we obtain

$$\left| \mathcal{R} \left(t, x \right) \right| = \left| \int_t^x \left(f^{\Delta} \left(\tau \right) - f^{\Delta} \left(x \right) \right) \Delta \tau \right| \leq \int_t^x \left| f^{\Delta} \left(\tau \right) - f^{\Delta} \left(x \right) \right| \Delta \tau$$

$$\leq M \int_t^x \left| \tau - x \right| \Delta \tau \leq M \left(x - t \right)^2. \tag{12.108}$$

So that, in either case, we have found

$$\left| \mathcal{R} \left(t, x \right) \right| \leq M \left(t - x \right)^2. \tag{12.109}$$

Hence we obtain

$$(L_n(f(t)))(x) = (L_n(f(x)))(x) + f^{\Delta}(x)(L_n(t-x))(x) + (L_n(\mathcal{R}(t,x)))(x), \tag{12.110}$$

which gives us

$$(L_n(f(t)))(x) - f(x) = f(x)((L_n(1))(x) - 1) \tag{12.111}$$
$$+ f^{\Delta}(x)(L_n(t-x))(x) + (L_n(\mathcal{R}(t,x)))(x).$$

Therefore it holds

$$|(L_n(f(t)))(x) - f(x)| \le |f(x)|\,|(L_n(1))(x) - 1| \tag{12.112}$$
$$+ \left|f^{\Delta}(x)\right| |(L_n(t-x))(x)| + (L_n(|\mathcal{R}(t,x)|))(x)$$
$$\le |f(x)|\,|(L_n(1))(x) - 1| + \left|f^{\Delta}(x)\right| |(L_n(t-x))(x)| + M\left(\left(L_n\left((t-x)^2\right)\right)(x)\right)$$
$$\le \|f\|_\infty \|L_n(1) - 1\|_\infty + \left\|f^{\Delta}\right\|_\infty \|(L_n(t-x))(x)\|_\infty + M\left\|\left(L_n\left((t-x)^2\right)\right)(x)\right\|_\infty. \tag{12.113}$$

Also, as earlier, we have

$$|(L_n(t-x))(x)| \le (L_n(|t-x|))(x)$$
$$\le \sqrt{(L_n(1))(x)}\sqrt{\left(L_n\left((t-x)^2\right)\right)(x)} \tag{12.114}$$
$$\le \sqrt{\|L_n(1)\|_\infty}\sqrt{\left\|\left(\left(L_n(t-x)^2\right)\right)(x)\right\|_\infty}.$$

That is

$$\|(L_n(t-x))(x)\|_\infty \le \sqrt{\|L_n(1)\|_\infty}\sqrt{\left\|\left(\left(L_n(t-x)^2\right)\right)(x)\right\|_\infty}. \tag{12.115}$$

The proof of the theorem now is complete. □

We mention

Theorem 12.13. *([4], [5], [9], Taylor's formula) Assume $\mathbb{T}^k = \mathbb{T}$ and $f \in C_{rd}^n(\mathbb{T})$ (the space of n times rd-continuously differentiable functions, see [5]), $n \in \mathbb{N}$ and $s, t \in \mathbb{T}$. Here generally define $h_0(t,s) = 1$, $\forall\ s,t \in \mathbb{T}$; $k \in \mathbb{N}_0$, and*

$$h_{k+1}(t,s) = \int_s^t h_k(\tau,s)\,\Delta\tau, \quad \forall\ s,t \in \mathbb{T}. \tag{12.116}$$

(then $h_k^{\Delta}(t,s) = h_{k-1}(t,s)$, for $k \in \mathbb{N}$, $\forall\ t \in \mathbb{T}$, for each $s \in \mathbb{T}$ fixed). Then

$$f(t) = \sum_{k=0}^{n-1} f^{\Delta^k}(s)\,h_k(t,s) + \int_s^t h_{n-1}(t,\sigma(\tau))\,f^{\Delta^n}(\tau)\,\Delta\tau \tag{12.117}$$

(above $f^{\Delta^0}(s) = f(s)$).

We need

Definition 12.8. ([5]) Let the functions $g_k : \mathbb{T}^2 \to \mathbb{R}$, $k \in \mathbb{N}_0$, defined recursively as follows:

$$g_0(t,s) = 1, \quad \forall\ s,t \in \mathbb{T}, \tag{12.118}$$

and

$$g_{k+1}(t,s) = \int_s^t g_k(\sigma(\tau), s)\,\Delta\tau, \quad \forall\ s,t \in \mathbb{T}. \tag{12.119}$$

Notice that

$$g_k^{\Delta}(t,s) = g_{k-1}(\sigma(t), s), \quad \text{for } k \in \mathbb{N},\ t \in \mathbb{T}^k. \tag{12.120}$$

Also it holds

$$g_1(t,s) = h_1(t,s) = t - s, \quad \forall\ s,t \in \mathbb{T}. \tag{12.121}$$

We need

Theorem 12.14. *([5]) It holds*

$$h_n(t,s) = (-1)^n g_n(s,t), \quad \forall\ t \in \mathbb{T} \tag{12.122}$$

and all $s \in \mathbb{T}^{k^n}$ *(*$\mathbb{T}^{k^2} := \left(\mathbb{T}^k\right)^k$*, ...,etc.)*

We need

Corollary 12.2. *Assume* $\mathbb{T}^k = \mathbb{T}$, $f \in C_{rd}^n(\mathbb{T})$, $n \in \mathbb{N}$ *and* $s,t \in \mathbb{T}$.
Then

$$f(t) = \sum_{k=0}^{n-1} f^{\Delta^k}(s)\,h_k(t,s)$$

$$+ (-1)^n f^{\Delta^n}(s)\,g_n(s,t) + \int_s^t h_{n-1}(t,\sigma(\tau))\left(f^{\Delta^n}(\tau) - f^{\Delta^n}(s)\right)\Delta\tau. \tag{12.123}$$

Proof. By (12.117) and (12.122). Namely we have

$$\int_s^t h_{n-1}(t,\sigma(\tau))\,\Delta\tau = (-1)^{n-1}\int_s^t g_{n-1}(\sigma(\tau),t)\,\Delta\tau \tag{12.124}$$

$$= (-1)^n \int_t^s g_{n-1}(\sigma(\tau),t)\,\Delta\tau = (-1)^n g_n(s,t).$$

□

We make

Remark 12.3. (to Corollary 12.2) One can easily prove inductively that

$$|h_k(t,s)| \le |t-s|^k, \quad \forall\ s,t \in \mathbb{T}, \tag{12.125}$$

$\forall\ k \in \mathbb{N}_0$.

Call

$$\mathcal{R}(t,s) := \int_s^t h_{n-1}(t,\sigma(\tau)) \left(f^{\Delta^n}(\tau) - f^{\Delta^n}(s)\right) \Delta\tau. \tag{12.126}$$

Assume that

$$\left|f^{\Delta^n}(t) - f^{\Delta^n}(s)\right| \leq M |t-s|^{\alpha}, \ \forall\, s,t \in \mathbb{T}, \tag{12.127}$$

where α is fixed such that $0 < \alpha \leq 1$.

We estimate $\mathcal{R}(t,s)$.

(i) Assume $t \geq s$. Then

$$|\mathcal{R}(t,s)| \leq \int_s^t |h_{n-1}(t,\sigma(\tau))| \left|f^{\Delta^n}(\tau) - f^{\Delta^n}(s)\right| \Delta\tau \tag{12.128}$$

$$\leq M \int_s^t |t-\sigma(\tau)|^{n-1} |\tau - s|^{\alpha} \Delta\tau \leq M (\sigma(t)-s)^{n-1} \int_s^t |\tau-s|^{\alpha} \Delta\tau$$

$$\leq M(\sigma(t)-s)^{n-1}(t-s)^{\alpha+1}. \tag{12.129}$$

We proved, if $t \geq s$ then

$$|\mathcal{R}(t,s)| \leq M(\sigma(t)-s)^{n-1}(t-s)^{\alpha+1} \leq M(\sigma(t)-s)^{n+\alpha}. \tag{12.130}$$

(ii) Assume $t < s$. Then

$$|\mathcal{R}(t,s)| \leq \int_t^s |h_{n-1}(t,\sigma(\tau))| \left|f^{\Delta^n}(\tau) - f^{\Delta^n}(s)\right| \Delta\tau \tag{12.131}$$

$$\leq M \int_t^s |t-\sigma(\tau)|^{n-1} |\tau-s|^{\alpha} \Delta\tau \leq M(\sigma(s)-t)^{n-1}(s-t)^{\alpha+1}$$

$$\leq M(\sigma(s)-t)^{n+\alpha}.$$

We proved, if $t < s$ then

$$|\mathcal{R}(t,s)| \leq M(\sigma(s)-t)^{n+\alpha}. \tag{12.132}$$

I.e. we have found

$$|\mathcal{R}(t,s)| \leq M\varphi(t,s), \tag{12.133}$$

where

$$\varphi(t,s) := \begin{cases} (\sigma(t)-s)^{n+\alpha}, & \text{if } t \geq s, \\ (\sigma(s)-t)^{n+\alpha}, & \text{if } t < s, \end{cases} \tag{12.134}$$

notice that $\varphi(\cdot,s) \in C_{rd}(\mathbb{T}), \ \forall\, s \in \mathbb{T}$.

We give the following related result

Theorem 12.15. *Assume* $\mathbb{T}^k = \mathbb{T}$, $\mathbb{T}$ *is a time scale and* $s, t \in [a, b]_{\mathbb{T}} \subset \mathbb{T}$.

Let $(L_N)_{N \in \mathbb{N}}$ *be a sequence of positive linear operators,* L_N *is mapping* $C_{rd}([a, b]_{\mathbb{T}})$ *into itself, such that* $L_N(1) = 1$. *Consider* $f \in C_u^n([a, b]_{\mathbb{T}})$, $n \in \mathbb{N}$, *such that*

$$\left| f^{\Delta^n}(t) - f^{\Delta^n}(s) \right| \leq M |t - s|^{\alpha}, \quad \forall\, t, s \in [a, b]_{\mathbb{T}}, \tag{12.135}$$

α *is fixed such that* $0 < \alpha \leq 1$, $M > 0$.

Let also $\varphi(t, s)$ *defined by (12.134).*

Then

(i)

$$\left| (L_N(f(t)))(s) - f(s) - \sum_{k=1}^{n-1} f^{\Delta^k}(s) (L_N(h_k(t, s)))(s) \right.$$

$$\left. + (-1)^{n+1} f^{\Delta^n}(s) (L_N(g_n(s, t)))(s) \right|$$

$$\leq M (L_N(\varphi(t, s)))(s), \quad \forall\, s \in [a, b]_{\mathbb{T}}, \tag{12.136}$$

(ii) additionally assume that $f^{\Delta^k}(s) = 0$, $k = 1, ..., n$, *for a fixed* $s \in [a, b]_{\mathbb{T}}$, *to derive*

$$|(L_N(f(t)))(s) - f(s)| \leq M (L_N(\varphi(t, s)))(s), \tag{12.137}$$

and

(iii)

$$|(L_N(f(t)))(s) - f(s)| \leq \sum_{k=1}^{n-1} \left| f^{\Delta^k}(s) \right| |(L_N(h_k(t, s)))(s)|$$

$$+ \left| f^{\Delta^n}(s) \right| |(L_N(g_n(s, t)))(s)| + M (L_N(\varphi(t, s)))(s), \quad \forall\, s \in [a, b]_{\mathbb{T}}. \tag{12.138}$$

If $(L_N(\varphi(t, s)))(s) \to 0$, *by (ii), we get*

$$(L_N(f(t)))(s) \to f(s), \quad \text{as } N \to +\infty.$$

Other useful convergence consequences follow by (i) and (iii).

Proof. By (12.123) we derive

$$(\Delta_N(f(t)))(s) := (L_N(f(t)))(s) - f(s) - \sum_{k=1}^{n-1} f^{\Delta^k}(s) (L_N(h_k(t, s)))(s)$$

$$+ (-1)^{n+1} f^{\Delta^n}(s) (L_N(g_n(s, t)))(s) = (L_N(\mathcal{R}(t, s)))(s). \tag{12.139}$$

Hence it holds

$$|(\Delta_N(f(t)))(s)| \leq (L_N(|\mathcal{R}(t, s)|))(s) \overset{(12.133)}{\leq} M (L_N(\varphi(t, s)))(s), \tag{12.140}$$

$\forall\, s \in [a, b]_{\mathbb{T}}$, proving the claim. □

Example 12.2 (Application). *Let $[a,b]_{\mathbb{T}} \subset \mathbb{T}$, $\mathbb{T}$ is a time scale, such that $\rho(b) = b$. Hence $C([a,b]_{\mathbb{T}}) = C_u([a,b]_{\mathbb{T}})$. Let $f \in C([a,b]_{\mathbb{T}})$, and $N \in \mathbb{N}$. Take $\varepsilon = \frac{1}{N}$, by uniform continuity of f on the compact set $[a,b]_{\mathbb{T}}$, there exists a $\delta = \delta(\varepsilon) > 0$ such that $t', t'' \in [a,b]_{\mathbb{T}}$, $|t' - t''| \leq \delta$ implies*

$$|f(t') - f(t'')| < \frac{1}{N}. \tag{12.141}$$

Define again

$$\chi_t(\xi) := \begin{cases} 1, & \text{for } \xi \in [a,t)_{\mathbb{T}}, \\ 0, & \text{for } \xi \in [t,b]_{\mathbb{T}}, \end{cases} \tag{12.142}$$

for $t \in [a,b)_{\mathbb{T}}$, and

$$\chi_b(\xi) = 1 \text{ for all } \xi \in [a,b]_{\mathbb{T}}. \tag{12.143}$$

Notice that $\chi_a(\xi) = 0$, $\xi \in [a,b]_{\mathbb{T}}$.

For a partition $P = \{t_0, t_1, ..., t_n\}$ ($t_0 = a < t_1 < ... < t_{n-1} < t_n = b$) belonging to $\mathcal{P}_\delta([a,b]_{\mathbb{T}})$ we consider the step function $f^{(\frac{1}{N})}$ defined on $[a,b]_{\mathbb{T}}$ by

$$f^{(\frac{1}{N})}(t) := f(t_{k-1}), \quad \text{if } t \in [t_{k-1}, t_k)_{\mathbb{T}},$$

$k = 1, ..., n$, and

$$f^{(\frac{1}{N})}(b) := f(t_n) = f(b),$$

which can be written in the form

$$f^{(\frac{1}{N})}(t) = \sum_{k=1}^{n} f(t_{k-1}) \left[\chi_{t_k}(t) - \chi_{t_{k-1}}(t)\right], \tag{12.144}$$

for all $t \in [a,b]_{\mathbb{T}}$.

In the proof of Theorem 12.5 we saw that for $t_{k-1} < t_k$ we get $0 \leq \chi_{t_{k-1}} \leq \chi_{t_k} \leq ... \leq \chi_b$. Thus it holds

$$\chi_{t_k}(t) - \chi_{t_{k-1}}(t) \geq 0, \ \forall\, t \in [a,b]_{\mathbb{T}}. \tag{12.145}$$

We define the following linear operator on $C([a,b]_{\mathbb{T}})$:

$$(L_N(f))(t) := f^{(\frac{1}{N})}(t) = \sum_{k=1}^{n} f(t_{k-1}) \left[\chi_{t_k}(t) - \chi_{t_{k-1}}(t)\right], \tag{12.146}$$

$\forall\, t \in [a,b]_{\mathbb{T}}$.

Clearly L_N is a positive linear operator from $C_u([a,b]_{\mathbb{T}})$ into $B([a,b]_{\mathbb{T}})$. Notice that $(L_n(f))(t_k) = f^{(\frac{1}{N})}(t_k) = f(t_k)$, for $k = 0, 1, ..., n$, that is $(L_N(f))$ has the interpolation property over P.

As in the proof of Theorem 12.5 we get

$$\|L_N f - f\|_\infty = \left\|f^{(\frac{1}{N})} - f\right\|_\infty \leq \frac{1}{N}. \tag{12.147}$$

The last proves uniform convergence of $L_N f$ to f, as $N \to +\infty$.

The above genuine example on time scales proves that our theory of approximation by positive linear operators on time scales is not trivial and not only valid on continuous intervals $[a,b]$ of $\mathbb{R}$.

Bibliography

1. G.A. Anastassiou, *Approximation by interpolating neural network operators*, Neural Parallel Scientific Computations, accepted 2014.
2. G.A. Anastassiou, *Approximation theory and functional analysis on time scales*, Internat. J. Difference Equations, accepted 2014.
3. H. Bauer, *Approximation and abstract boundaries*, Ann. Math. Monthly 85 (1978), 632-647.
4. M. Bohner and G. Guseinov, *The convolution on time scales*, Abstract Appl. Anal. 2007 (2007), Article ID 58373, 24 pages.
5. M. Bohner and A. Peterson, *Dynamic Equations on Time Scales: An Introduction with Applications*, Birkhäuser, Boston, 2001.
6. M. Bohner and A. Peterson, Editors, *Advances in Dynamic Equations on Time Scales*, Birkhäuser, Boston, Berlin, 2003.
7. K. Boulabiar, *A Hölder-type inequality for positive functionals on Φ-Algebras*, J. Inequalities Pure Appl. Math. 3 (5) (2002), Article 74.
8. G.Sh. Guseinov, *Integration on time scales*, J. Math. Anal. Appl. 285 (2003), 107-127.
9. R. Higgins and A. Peterson, *Cauchy functions and Taylor's formula for Time scales T*, (2004), in *Proc. Sixth. Internat. Conf. on Difference Equations*, edited by B. Aulbach, S. Elaydi, G. Ladas, pp. 299-308, New Progress in Difference Equations, Augsburg, Germany, 2001, Chapman & Hall / CRC.
10. A. Huseynov, *The Riesz representation theorem on time scales*, Math. Comput. Modell. 55 (2012), 1570-1579.
11. P.P. Korovkin, *Linear Operators and Approximation Theory*, Hindustan Publ. Corp., Delhi, India, 1960.
12. P. Lax, *Functional Analysis*, John Wiley & Sons, New York, 2002.
13. O. Shisha and B. Mond, *The degree of convergence of sequences of linear positive operators*, Proc. Natl. Acad. Sci. USA 60 (1968), 1196-1200.

Index

Printed in the United States
By Bookmasters